ROWAN UNIVERSITY
CAMPBELL LIBRARY
201 MULLICA HILL RD.
GLASSBORO, NJ 08028-1701

Elicitation of Expert Opinions for Uncertainty and Risks

Elicitation of Expert Opinions for Uncertainty and Risks

Bilal M. Ayyub

CRC Press
Boca Raton   London   New York   Washington, D.C.

Library of Congress Cataloging-in-Publication Data

Ayyub, Bilal M.
  Elicitation of expert opinions for uncertainty and risks by Bilal M. Ayyub
    p. cm.
  Includes bibliographical references and index.
  ISBN 0-8493-1087-3 (alk. paper)
  1. Industrial engineering.
  2. System analysis.
  3. Decision making. I. Title.
  T56 .A98 2001
  658.4'6.—dc21

2001025644

This book contains information obtained from authentic and highly regarded sources. Reprinted material is quoted with permission, and sources are indicated. A wide variety of references are listed. Reasonable efforts have been made to publish reliable data and information, but the author and the publisher cannot assume responsibility for the validity of all materials or for the consequences of their use.

Neither this book nor any part may be reproduced or transmitted in any form or by any means, electronic or mechanical, including photocopying, microfilming, and recording, or by any information storage or retrieval system, without prior permission in writing from the publisher.

The consent of CRC Press LLC does not extend to copying for general distribution, for promotion, for creating new works, or for resale. Specific permission must be obtained in writing from CRC Press LLC for such copying.

Direct all inquiries to CRC Press LLC, 2000 N.W. Corporate Blvd., Boca Raton, Florida 33431.

Trademark Notice: Product or corporate names may be trademarks or registered trademarks, and are used only for identification and explanation, without intent to infringe.

Visit the CRC Press Web site at www.crcpress.com

© 2001 by CRC Press LLC

No claim to original U.S. Government works
International Standard Book Number 0-8493-1087-3
Library of Congress Card Number 2001025644
Printed in the United States of America 1 2 3 4 5 6 7 8 9 0
Printed on acid-free paper

Dedication

To my wife, Deena, and our children, Omar, Rami, Samar, and Ziad

Preface

The complexity of our society and its knowledge base requires its members to specialize and become *experts* to attain recognition and reap rewards for society and themselves. We commonly deal with or listen to experts on a regular basis, such as weather forecasts by weather experts, stock and or financial reports by analysts and experts, suggested medication or procedures by medical professionals, policies by politicians, or analyses by world-affairs experts. We know from our own experiences that experts are valuable sources of information and knowledge, but that they can also be wrong in their views. Expert opinions, therefore, can be considered to include or constitute non-factual information. The fallacy of these opinions might disappoint us, but does not surprise us, since issues that require experts tend to be difficult or complex, sometimes with divergent views. The nature of some of these complex issues could yield only views that have subjective truth levels; therefore, they allow for contradictory views that might all be somewhat credible. In political and economic world affairs and international conflicts, such issues are common. For example, we have witnessed the debates that surrounded the membership of the People's Republic of China to the World Trade Organization in 1999, experts airing their views on the Arab-Israeli affairs in 2000, analysts' views on the 1990 sanctions on the Iraqi people, and future oil prices. Also, such issues and expert opinions are common in engineering, sciences, medical fields, social research, stock and financial markets, and the legal practice.

Experts, with all their importance and value, can be viewed as double-edged swords. Not only do they bring a deep knowledge base and thoughts, but they also could provide biases and *pet* theories. The selection of experts, elicitation of their opinions, and aggregating the opinions should be performed and handled carefully by recognizing uncertainties associated with those opinions, and sometimes skeptically.

The primary reason for eliciting expert-opinion is to deal with uncertainty in selected technical issues related to a system of interest. Issues with significant uncertainty, issues that are controversial and/or contentious, issues that are complex, and/or issues that can have a significant effect on risk are most suited for expert-opinion elicitation. The value of the expert-opinion comes from its initial intended uses as a heuristic tool, not as a scientific tool, for exploring vague and unknowable issues that are otherwise inaccessible. It is not a substitute for rigorous, scientific research.

In preparing this book, I strove to achieve the following objectives: (1) develop a philosophical foundation for the meaning, nature, and hierarchy of knowledge and ignorance; (2) provide background information and historical developments related to knowledge, ignorance, and the elicitation of expert opinions; (3) provide methods for expressing expert opinions and aggregating them; (4) guide the readers of the book on how to effectively elicit opinions of experts that would increase the truthfulness of the outcomes of an expert-opinion elicitation process; and (5) provide practical applications based on recent elicitations that I facilitated. In covering methods for expressing expert opinions and aggregating them, the book introduces relevant, fundamental concepts of classical sets, fuzzy sets, rough sets, probability, Bayesian methods, interval analysis, fuzzy arithmetic, interval probabilities, evidence theory, and possibility theory. These methods are presented in a style tailored to meet the needs of engineering, sciences, economics, and law students and practitioners. The book emphasizes the practical use of these methods, and establishes the limitations, advantages, and disadvantages of the methods. Although, the applications at the end of the book were developed with emphasis on engineering, technological, and economic problems, the methods can also be used to solve problems in other fields, such as the sciences, law, and management.

Problems that are commonly encountered by engineers and scientists require decision-making under conditions of uncertainty, lack of knowledge, and ignorance. The lack of knowledge and ignorance can be related to the definition of a problem, the alternative solution methodologies and their results, and the nature of the solution outcomes. Studies show that in the future, analysts, engineers, and scientists will need to solve more complex problems with decisions made under conditions of limited resources, thus necessitating increased reliance on the proper treatment of uncertainty and the use of expert opinions. Therefore, this book is intended to better prepare future analysts, engineers, and scientists, as well as assist practitioners in understanding the fundamentals of knowledge and ignorance, how to elicit expert opinions, how to select appropriate expressions of these opinions, and how to aggregate the opinions. Also, the book is intended to better prepare them to use appropriately and adequately various methods for modeling and aggregating expert opinions.

Structure, format, and main features

This book was written with a dual use in mind, as both a self-learning guidebook and as a required textbook for a course. In either case, the text has been designed to achieve important educational objectives of introducing theoretical bases, guidance, and applications of expert-opinion elicitation.

The seven chapters of the book lead the readers from the definition of needs, to foundations of the concepts covered in the book, to theory, and finally to guidance and applications. The first chapter provides an introduction to the book by discussing knowledge, its sources and acquisition, and ignorance and its categories as bases for dealing with experts and their opinions. The practical use of concepts and tools presented in the book requires a framework and a frame of thinking that deals holistically with problems and issues as systems. Background information on system modeling is provided in Chapter 2. Chapter 3 provides background information on experts, opinions, expert-opinion elicitation methods, methods used in developing questionnaires in educational and psychological testing and social research, and methods and practices utilized in focus groups. Chapter 4 presents the fundamentals of classical set theory, fuzzy sets, and rough sets that can be used to express opinions. Basic operations for these sets are defined and demonstrated. Fuzzy relations and fuzzy arithmetic can be used to express and combine information collected. The fundamentals of probability theory, possibility theory, interval probabilities, and monotone measures are summarized as they relate to the expression of expert opinions. Examples are used in this chapter to demonstrate the various methods and concepts. Chapter 5 presents methods for assessing or scoring expert opinions, measuring uncertainty contents in individual opinions and aggregated or combined opinions, and selecting an optimal opinion. The methods presented in Chapter 5 are based on developments in expert-opinion elicitation and uncertainty-based information in the field of information science. Chapter 6 provides guidance on using expert-opinion elicitation processes. These processes can be viewed as variations of the Delphi technique with scenario analysis based on uncertainty models, ignorance, knowledge, information, and uncertainty modeling related to experts and opinions and nuclear industry experiences and recommendations. Chapter 7 demonstrates the applications of expert-opinion elicitation by focusing on occurrence probabilities and consequences of events related to naval and civil works systems for the purposes of planners, engineers, and others, who may use expert opinions.

In each chapter of the book, computational examples are given in the individual sections of the chapter, with more detailed engineering applications given in a concluding chapter. Also, each chapter includes a set of exercise problems that cover the materials of the chapter. The problems were carefully designed to meet the needs of instructors in assigning homework and of readers in practicing the fundamental concepts.

For the purposes of teaching, the book can be covered in one semester. The chapter sequence can be followed as a recommended sequence. However, if needed, instructors can choose a subset of the chapters for courses that do not permit a complete coverage of all chapters or a coverage that cannot follow the order presented. In addition, selected chapters can be used to supplement courses that do not deal directly with expert-opinion elicitation, such as risk analysis, reliability assessment, economic analysis, systems analysis, legal opinions, and social research courses. Chapters 1, 2, and 3 can be covered concurrently. Chapter 4 builds on some of the material covered in Chapter 3. Chapter 5 builds on Chapters 3 and 4, and should be covered after completing Chapter 4. Chapter 6 provides guidance on using expert-opinion elicitation, and can be introduced after the preceding chapters. Chapter 7 provides applications. The book also contains an extensive reference section at its end. The accompanying schematic diagram illustrates possible sequences of these chapters in terms of their interdependencies.

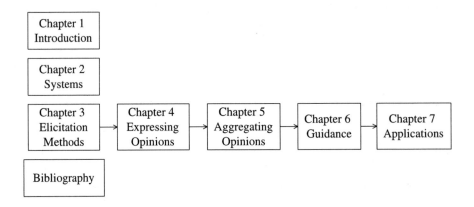

Acknowledgments

This book was developed over several years and draws on my experiences in teaching courses related to risk analysis, uncertainty modeling and analysis, probability and statistics, numerical methods and mathematics, reliability assessment, and decision analysis. Drafts of most sections of the book were tested in several courses at the University of Maryland, College Park, for about three years before its publication. This testing period has proved to be a very valuable tool in establishing its contents and the final format and structure.

I was very fortunate to receive the direct and indirect help from many individuals, over many years, that greatly affected this book. Students who took courses and used portions of this book provided me with great insight on how to effectively communicate various theoretical concepts. Also, students' research projects and my interaction with them stimulated the generation of various examples. The students who took courses on structural reliability, risk analysis, and mathematical methods in civil engineering in the 1990s contributed to this endeavor. Their feedback was very helpful and greatly contributed to the final product. Also, comments provided by M. Al-Fadhala, I. Assakkaf, M. Kaminskiy, and R. Wlicox on selected chapters are greatly appreciated.

I was fortunate to organize the International Symposia on Uncertainty Modeling and Analysis in 1990, 1993, and 1995. These symposia were tremendously useful, as they provided me with rewarding opportunities to meet, interact, and learn from researchers and scientists from more than 35 countries, including most notably Professors D. Blockley, C.B. Brown, H. Furuta, M. Gupta, A. Haldar, L. Kanal, A. Kaufmann, G.J. Klir, R.R. Yager, J.T.P. Yao, L.A. Zadeh, H.G. Zimmerman.

The reviewers' comments that were provided by the publisher were used to improve the book to meet the needs of readers and enhance the educational process. The input from the publisher and the book reviewers is greatly appreciated.

The financial support that I received from the U.S. Navy, Coast Guard, Army Corps of Engineers, National Science Foundation, and the American Society of Mechanical Engineers over more than 15 years has contributed greatly to this book by providing me with a wealth of information and ideas for formalizing the theory, applications, and guidance. In particular,

I acknowledge the opportunity and support provided by A. Ang, R. Art, K. Balkey, J. Beach, P. Capple, J. Crisp, D. Dressler, M. Firebaugh, J. Foster, Z. Karaszewski, D. Moser, G. Remmers, T. Shugar, S. Taylor, S. Wehr, and G. White.

The University of Maryland at College Park has provided me with the platform, support, and freedom that made such a project possible. It has always provided me with a medium for creativity and excellence. I am indebted all my life for what the University of Maryland at College Park, especially the A. James Clark School of Engineering and the Department of Civil and Environmental Engineering, has done for me. The students, staff, and my colleagues define this fine institution and its units.

Last but not least, I am boundlessly grateful to my family for accepting my absences, sometimes physical and sometimes mental, as I worked on projects related to this book and its pages, for making it possible, and for making my days worthwhile. I also greatly appreciate the boundless support of my parents, brothers, and sisters: Thuraya, Mohammed, Saleh, Naser, Nidal, Intisar, Jamilah, and Mai. Finally, one individual who has accepted my preoccupation with the book and absence, kept our life on track, and filled our life with treasures — or all of that, thanks to my wife, Deena; none of this would be possible without her.

I invite users of the book to send any comments on the book to the e-mail address **ayyub@umail.umd.edu**. These comments will be used in developing future editions of the book. Also, I invite users of the book to visit the web site of the Center for Technology and Systems Management at the University of Maryland, College Park, to find information posted on various projects and publications that can be related to expert-opinion elicitation. The URL is **http://ctsm.umd.edu**.

Bilal M. Ayyub

About the author

Bilal M. Ayyub is a professor of civil and environmental engineering at the University of Maryland (College Park) and the General Director of the Center for Technology and Systems Management. He is also a researcher and consultant in the areas of structural engineering, systems engineering, uncertainty modeling and analysis, reliability and risk analysis, and applications related to civil, marine, and mechanical systems. He completed his B.S. degree in civil engineering in 1980, and completed both the M.S. (1981) and Ph.D. (1983) degrees in civil engineering at the Georgia Institute of Technology. He has performed several research projects that were funded by the U.S. National Science Foundation, Coast Guard, Navy, Army Corps of Engineers, Maryland State Highway Administration, American Society of Mechanical Engineers, and several engineering companies. Dr. Ayyub served the engineering community in various capacities through societies that include ASNE, ASCE, ASME, SNAME, IEEE, and NAFIPS. He is a fellow of ASCE, ASME, and SNAME, and life member of ASNE and USNI. He chaired the ASCE Committee on the Reliability of Offshore Structures, and currently chairs the SNAME panel on design philosophy and the ASNE Naval Engineers Journal committee. He also was the General Chairman of the first, second and third International Symposia on Uncertainty Modeling and Analysis that were held in 1990, 1993, and 1995, and NAFIPS annual conference in 1995. He is the author and coauthor of approximately 300 publications in journals, and conference proceedings, and reports. His publications include several textbooks and edited books. Dr. Ayyub is the triple recipient of the ASNE "Jimmie" Hamilton Award for the best papers in the *Naval Engineers Journal* in 1985, 1992, and 2000. Also, he received the ASCE award for "Outstanding Research Oriented Paper" in the *Journal of Water Resources Planning and Management* for 1987, the ASCE Edmund Friedman Award in 1989, and in 1997 the NAFIPS K.S. Fu Award for distinguished service and Walter L. Huber Research Prize. He is a registered Professional Engineer (PE) with the state of Maryland. He is listed in *Who's Who in America*, and *Who's Who in the World*.

Books by Bilal M. Ayyub

Probability, Statistics and Reliability for Engineers, CRC Press, 1997, B.M. Ayyub and R. McCuen.

Numerical Methods for Engineers, Prentice Hall, New York, 1996, by B.M. Ayyub and R. McCuen.

Uncertainty Modeling and Analysis in Civil Engineering, CRC Press, 1998, by B.M. Ayyub (Editor).

Uncertainty Modeling in Vibration, Control, and Fuzzy Analysis of Structural Systems, World Scientific, 1997, by B.M. Ayyub, A. Guran, and A. Haldar (Editors).

Uncertainty Analysis in Engineering and the Sciences: Fuzzy Logic, Statistics, and Neural Network Approach, Kluwer Academic Publisher, 1997, by B.M. Ayyub and M.M. Gupta (Editors).

Uncertainty Modeling in Finite Element, Fatigue, and Stability of Systems, World Scientific, 1997, by A. Haldar, A. Guran, and B.M. Ayyub (Editors).

Uncertainty Modeling and Analysis: Theory and Applications, North-Holland-Elsevier Scientific Publishers, 1994, by B.M. Ayyub and M.M. Gupta (Editors).

Analysis and Management of Uncertainty: Theory and Applications, North-Holland-Elsevier Scientific Publishers, by B.M. Ayyub and M.M. Gupta (Editors).

Contents

Chapter 1. Knowledge and ignorance
1.1. Information abundance and ignorance 1
1.2. The nature of knowledge 3
    1.2.1. Basic terminology and definitions 3
    1.2.2. Absolute reality and absolute knowledge 3
    1.2.3. Historical developments and perspectives 6
        1.2.3.1. The preSocratic period 6
        1.2.3.2. The Socratic period 6
        1.2.3.3. The Plato and Aristotle period 7
        1.2.3.4. The Hellenistic 10
        1.2.3.5. The Medieval 10
        1.2.3.6. The Renaissance 12
        1.2.3.7. The 17th century 13
        1.2.3.8. The 18th century 14
        1.2.3.9. The 19th century 15
        1.2.3.10. The 20th century 16
    1.2.4. Knowledge, information, and opinions 18
1.3. Cognition and cognitive science 22
1.4. Time and its asymmetry 25
1.5. Defining ignorance in the context of knowledge 26
    1.5.1. Human knowledge and ignorance 26
    1.5.2. Classifying ignorance 28
    1.5.3. Ignorance hierarchy 31
1.6. Exercise problems 34

Chapter 2. Information-based system definition
2.1. Introduction 38
2.2. System definition models 42
    2.2.1. Perspectives for system definition 42
    2.2.2. Requirements and work breakdown structure 43
        2.2.2.1. Requirements analysis 43
        2.2.2.2. Work breakdown structure 45
    2.2.3. Process modeling method 45
        2.2.3.1. System engineering process 46

| | | 2.2.3.2. | Lifecycle of engineering systems .... 49 |
| | | 2.2.3.3. | Technical maturity model .... 54 |

- 2.2.4. Black-box method 54
- 2.2.5. State-based method 57
- 2.2.6. Component integration method 58
- 2.2.7. Decision analysis method 59
  - 2.2.7.1. Decision variables 59
  - 2.2.7.2. Decision outcomes 61
  - 2.2.7.3. Associated probabilities and consequences 61
  - 2.2.7.4. Decision trees 61
  - 2.2.7.5. Influence diagrams 64
- 2.3. Hierarchical definitions of systems 64
  - 2.3.1. Introduction 64
  - 2.3.2. Knowledge and information hierarchy 66
    - 2.3.2.1. Source systems 67
    - 2.3.2.2. Data systems 72
    - 2.3.2.3. Generative systems 75
    - 2.3.2.4. Structure systems 81
    - 2.3.2.5. Metasystems 82
- 2.4. Models for ignorance and uncertainty types 83
  - 2.4.1. Mathematical theories for ignorance types 83
  - 2.4.2. Information uncertainty in engineering systems 85
    - 2.4.2.1. Abstraction and modeling of engineering systems 85
    - 2.4.2.2. Ignorance and uncertainty in abstracted aspects of a system 88
    - 2.4.2.3. Ignorance and uncertainty in nonabstracted aspects of a system 90
    - 2.4.2.4. Ignorance due to unknown aspects of a system 90
- 2.5. System complexity 91
- 2.6. Exercise problems 95

Chapter 3. Experts, opinions, and elicitation methods

- 3.1. Introduction 97
- 3.2. Experts and expert opinions 98
- 3.3. Historical background 99
  - 3.3.1. Delphi method 99
  - 3.3.2. Scenario analysis 105
- 3.4. Scientific heuristics 105
- 3.5. Rational consensus 109
- 3.6. Elicitation methods 111
  - 3.6.1. Indirect elicitation 111
  - 3.6.2. Direct method 113
  - 3.6.3. Parametric estimation 114
- 3.7. Standards for educational and psychological testing 114

- 3.8. Methods of social research 118
- 3.9. Focus groups 123
- 3.10. Exercise problems 124

Chapter 4. Expressing and modeling expert opinions
- 4.1. Introduction 126
- 4.2. Set theory 127
  - 4.2.1. Sets and events 127
  - 4.2.2. Fundamentals of classical set theory 127
    - 4.2.2.1. Classifications of sets 127
    - 4.2.2.2. Subsets 128
    - 4.2.2.3. Membership (or characteristic) function 128
    - 4.2.2.4. Sample space and events 128
    - 4.2.2.5. Venn-Euler diagrams 129
    - 4.2.2.6. Basic operations on sets 130
  - 4.2.3. Fundamentals of fuzzy sets and operations 131
    - 4.2.3.1. Membership (or characteristic) function 132
    - 4.2.3.2. Alpha-cut sets 134
    - 4.2.3.3. Fuzzy Venn-Euler diagrams 135
    - 4.2.3.4. Fuzzy numbers, intervals and arithmetic 137
    - 4.2.3.5. Operations on fuzzy sets 145
    - 4.2.3.6. Fuzzy relations 152
    - 4.2.3.7. Fuzzy functions 155
  - 4.2.4. Fundamental of rough sets 156
    - 4.2.4.1. Rough set definitions 156
    - 4.2.4.2. Rough set operations 157
    - 4.2.4.3. Rough membership functions 159
    - 4.2.4.4. Rough functions 160
- 4.3. Monotone measures 162
  - 4.3.1. Definition of monotone measures 163
  - 4.3.2. Classifying monotone measures 164
  - 4.3.3. Evidence theory 165
    - 4.3.3.1. Belief measure 165
    - 4.3.3.2. Plausibility measure 166
    - 4.3.3.3. Basic assignment 166
  - 4.3.4. Probability theory 170
    - 4.3.4.1. Relationship between evidence theory and probability theory 170
    - 4.3.4.2. Classical definitions of probability 171
    - 4.3.4.3. Linguistic probabilities 173
    - 4.3.4.4. Failure rates 173
    - 4.3.4.5. Central tendency measures 173
    - 4.3.4.6. Dispersion (or variability) 176
    - 4.3.4.7. Percentiles 177
    - 4.3.4.8. Statistical uncertainty 178
    - 4.3.4.9. Bayesian methods 181

		4.3.4.10. Interval probabilities 188
		4.3.4.11. Interval cumulative distribution functions 189
		4.3.4.12. Probability bounds 193
	4.3.5.	Possibility theory 199
4.4.	Exercise problems 201

Chapter 5. Consensus and aggregating expert opinions
5.1.	Introduction 208
5.2.	Methods of scoring of expert opinions 208
	5.2.1.	Self scoring 208
	5.2.2.	Collective scoring 209
5.3.	Uncertainty measures 209
	5.3.1.	Types of uncertainty measures 209
	5.3.2.	Nonspecificity measures 209
	5.3.3.	Entropy-like measures 211
		5.3.3.1. Shannon entropy for probability theory 212
		5.3.3.2. Discrepancy measure 212
		5.3.3.3. Entropy measures for evidence theory 213
	5.3.4.	Fuzziness measure 214
	5.3.5.	Other measures 214
5.4.	Combining expert opinions 214
	5.4.1.	Consensus combination of opinions 215
	5.4.2.	Percentiles for combining opinions 215
	5.4.3.	Weighted combinations of opinions 215
	5.4.4.	Uncertainty-based criteria for combining expert opinions 218
		5.4.4.1. Minimum uncertainty criterion 219
		5.4.4.2. Maximum uncertainty criterion 219
		5.4.4.3. Uncertainty invariance criterion 221
	5.4.5.	Opinion aggregation using interval analysis and fuzzy arithmetic 221
	5.4.6.	Opinion aggregation using Dempster's rule of combination 222
	5.4.7.	Demonstrative examples of aggregating expert opinions 222
		5.4.7.1. Aggregation of expert opinions 222
		5.4.7.2. Failure classification 227
5.5.	Exercise problems 230

Chapter 6. Guidance on expert-opinion elicitation
6.1.	Introduction and terminology 233
	6.1.1.	Theoretical bases 233
	6.1.2.	Terminology 234
6.2.	Classification of issues, study levels, experts, and process outcomes 236
6.3.	Process definition 238

- 6.4. Need identification for expert-opinion elicitation 239
- 6.5. Selection of study level and study leader 239
- 6.6. Selection of peer reviewers and experts 241
  - 6.6.1. Selection of peer reviewers 241
  - 6.6.2. Identification and selection of experts 241
  - 6.6.3. Items needed by experts and reviewers before the expert-opinion elicitation meeting 243
- 6.7. Identification, selection, and development of technical issues 244
- 6.8. Elicitation of opinions 245
  - 6.8.1. Issue familiarization of experts 245
  - 6.8.2. Training of experts 245
  - 6.8.3. Elicitation and collection of opinions 246
  - 6.8.4. Aggregation and presentation of results 247
  - 6.8.5. Group interaction, discussion, and revision by experts 247
- 6.9. Documentation and communication 247
- 6.10. Exercise problems 247

Chapter 7. Applications of expert-opinion elicitation
- 7.1. Introduction 249
- 7.2. Assessment of occurrence probabilities 250
  - 7.2.1. Cargo elevators onboard ships 250
    - 7.2.1.1. Background 250
    - 7.2.1.2. Example issues and results 250
  - 7.2.2. Navigation locks 251
    - 7.2.2.1. Background 251
    - 7.2.2.2. General description of lock operations 251
    - 7.2.2.3. Description of components 255
    - 7.2.2.4. Example issues and results 257
- 7.3. Economic consequences of floods 262
  - 7.3.1. Background 262
  - 7.3.2. The Feather River Basin 262
    - 7.3.2.1. Levee failure and consequent flooding 262
    - 7.3.2.2. Flood characteristics 262
    - 7.3.2.3. Building characteristics 263
    - 7.3.2.4. Vehicle characteristics 264
  - 7.3.3. Example issues and results 264
    - 7.3.3.1. Structural depth-damage relationships 264
    - 7.3.3.2. Content depth-damage relationships 264
    - 7.3.3.3. Content-to-structure value ratios 266
    - 7.3.3.4. Vehicle depth-damage relationship 266

Bibliography 279

Index 289

chapter one

Knowledge and ignorance

Contents

1.1. Information abundance and ignorance 1
1.2. The nature of knowledge 3
    1.2.1. Basic terminology and definitions 3
    1.2.2. Absolute reality and absolute knowledge 3
    1.2.3. Historical developments and perspectives 6
        1.2.3.1. The preSocratic period 6
        1.2.3.2. The Socratic period 6
        1.2.3.3. The Plato and Aristotle period 7
        1.2.3.4. The Hellenistic period 10
        1.2.3.5. The Medieval period 10
        1.2.3.6. The Renaissance 12
        1.2.3.7. The 17th century 13
        1.2.3.8. The 18th century 14
        1.2.3.9. The 19th century 15
        1.2.3.10. The 20th century 16
    1.2.4. Knowledge, information, and opinions 18
1.3. Cognition and cognitive science 22
1.4. Time and its asymmetry 25
1.5. Defining ignorance in the context of knowledge 26
    1.5.1. Human knowledge and ignorance 26
    1.5.2. Classifying ignorance 28
    1.5.3. Ignorance hierarchy 31
1.6. Exercise Problems 34

1.1 *Information abundance and ignorance*

Citizens of modern information-based, industrial societies are becoming increasingly aware of, and sensitive to, the harsh and discomforting reality that information abundance does not necessarily give us certainty. Sometimes it can lead to errors in decision-making with undesirable outcomes due to either overwhelming and confusing situations, or a sense of overconfidence leading to improper information use. The former situation can be an outcome of the limited capacity of a human mind in some situations to deal with complexity and information abundance; whereas the latter can be attributed to a higher order of ignorance, called the ignorance of self-ignorance.

As our society advances in many scientific dimensions and invents new technologies, human knowledge is being expanded through observation, discovery, information gathering, and logic. Also, the access to newly generated information is becoming easier than ever as a result of computers and the Internet. We are entering an exciting era where electronic libraries, online databases, and information on every aspect of our civilization, such as patents, engineering products, literature, mathematics, physics, medicine, philosophy, and public opinions will become a few mouse-clicks away. In this era, computers can generate even more information from abundantly available online information. Society can act or react based on this information at the speed of its generation, creating sometimes nondesirable situations, for example, price and/or political volatilities. There is a great need to assess uncertainties associated with information and to quantify our state of knowledge and/or ignorance. The accuracy, quality, and incorrectness of such information and knowledge incoherence are coming under focus by our philosophers, scientists, engineers, technologists, decision and policy makers, regulators and lawmakers, and our society as a whole. As a result, uncertainty and ignorance analyses are receiving a lot of attention by our society. We are moving from emphasizing the state of knowledge expansion and creation of information to a state that includes knowledge and information assessment by critically evaluating them in terms of relevance, completeness, nondistortion, coherence, and other key measures.

Our society is becoming less forgiving of, and more demanding from, our knowledge base. Untimely processing and use of any available information, even if it might be inconclusive, is considered worse than a lack of knowledge and ignorance. In 2000, the U.S. Congress and the Justice Department investigated Firestone and Ford Companies for allegedly knowing about their defective tires, suspected of causing accidents claiming more than 88 lives worldwide, without taking appropriate actions. The investigation and news elevated the problem to the level of scandal because of the company's inaction on available information, although the Firestone and Ford Companies argued that test results conducted, after they knew about a potential problem, were inconclusive. Such reasoning can easily be taken by our demanding society as a cover-up, causing a belligerent attitude that is even worse than the perception of inaction by corporate executives. Although people have some control over the levels of technology-caused risks to which they are exposed, governments and corporations need to pursue risk reduction as a result of increasing demands by our society, which generally entails a reduction of benefits, thus posing a serious dilemma. Policy makers and the public are required, with increasing frequency, to subjectively weigh benefits against risks and assess associated uncertainties when making decisions. Further, lacking a systems or holistic approach, vulnerability exists for overpaying to reduce one set of risks that may introduce offsetting or larger risks of another kind.

The objective of this chapter is to discuss knowledge, its sources, and acquisition, as well as ignorance and its categories as bases for dealing with experts and their opinions. The practical use of concepts and tools presented in the book requires a framework and a frame of thinking that deals holistically with problems and issues as systems. Background information on system modeling is provided in Chapter 2.

1.2 The nature of knowledge

1.2.1 Basic terminology and definitions

Philosophers have concerned themselves with the study of knowledge, truth, reality, and knowledge acquisition since the early days of Greece, including Thales (c. 585 BC), Anaximander (611–547 BC), and Anaximenes (c. 550 BC) who first proposed a rational explanation of the natural world and its powers. This section provides a philosophical introduction to knowledge, epistemology, their development, and related terminology.

Philosophy (philosophia) is a Greek term meaning *love of wisdom*. It deals with the careful thought about the fundamental nature of the world, the grounds for human knowledge, and the evaluation of human conduct. Philosophy, as an academic discipline, has chief branches that include logic, metaphysics, epistemology, and ethics. Selected terms related to knowledge and epistemology are defined in Table 1.1.

Philosophers' definitions of knowledge, its nature, and methods of acquisitions evolved over time, producing various schools of thought. In subsequent sections, these developments are briefly summarized along a historical timeline referring only to what was subjectively assessed as primary departures from previous schools. The new schools can be considered as new alternatives, since in some cases they could not invalidate previous ones.

1.2.2 Absolute reality and absolute knowledge

The absolute reality of things is investigated in a branch of philosophy called *metaphysics* that is concerned with providing a comprehensive account of the most general features of reality as a whole. The term metaphysics is believed to have originated in Rome about 70 BC by the Greek peripatetic philosopher Andronicus of Rhodes in his edition of the works of Aristotle (384–322 BC).

Metaphysics typically deals with issues related to the ultimate nature of things, identification of objects that actually exist, things that compose the universe, the ultimate reality, the nature of mind and substance, and the most general features of reality as a whole. On the other hand, *epistemology* is a branch of philosophy that investigates the possibility, origins, nature, and extent of human knowledge. Metaphysics and epistemology are very

Table 1.1 Selected Knowledge and Epistemology Terms

Term	Definition
Philosophy	The fundamental nature of the world, the grounds for human knowledge, and the evaluation of human conduct.
Epistemology	A branch of philosophy that investigates the possibility, origins, nature, and extent of human knowledge.
Metaphysics	The investigation of ultimate reality. A branch of philosophy concerned with providing a comprehensive account of the most general features of reality as a whole, and the study of being as such. Questions about the existence and nature of minds, bodies, God, space, time, causality, unity, identity, and the world are all metaphysical issues.
Ontology	A branch of metaphysics concerned with identifying, in the most general terms, the kinds of things that actually exist.
Cosmology	A branch of metaphysics concerned with the origin of the world.
Cosmogony	A branch of metaphysics concerned with the evolution of the universe.
Ethics	A branch of philosophy concerned with the evaluation of human conduct.
Aesthetics	A branch of philosophy that studies beauty and taste, including their specific manifestations in the tragic, the comic, and the sublime; where beauty is the characteristic feature of things that arouse pleasure or delight, especially to the senses of a human observer, and sublime is the aesthetic feeling aroused by experiences too overwhelming (i.e., awe) in scale to be appreciated as beautiful by the senses.
Knowledge	A body of propositions that meet the conditions of justified true belief.
Priori	Knowledge derived from reason alone.
Posteriori	Knowledge gained from intuitions and experiences.
Rationalism	Inquiry based on priori principles, or knowledge based on reason.
Empiricism	Inquiry based on posteriori principles, or knowledge based on experience.

closely linked and, at times, indistinguishable as the former speculates about the nature of reality, and latter speculates about the knowledge of it. Metaphysics is often formulated in terms of three modes of reality — the *immaterial mind* (or consciousness), the *matter* (or physical substance), and a *higher nature* (one which transcends both mind and matter) — according to three specific philosophical schools of thought: idealism, materialism, and transcendentalism, respectively.

Idealism is based on a theory of reality, derived from Plato's *Theory of Ideas* (427–347 BC) that attributes to *consciousness*, or the immaterial mind, a primary role in the constitution of the world. Metaphysical idealism contends that reality is mind-dependent and that true knowledge of reality is gained by relying upon a spiritual or conscious source.

The school of *materialism* is based on the notion that all existence is resolvable into matter, or into an attribute or effect of matter. Accordingly, matter is the ultimate reality, and the phenomenon of consciousness is explained by physiochemical changes in the nervous system. In metaphysics, materialism is the antithesis of idealism in which the supremacy of mind is affirmed, and matter is characterized as an aspect or objectification of mind. The world is considered to be entirely mind-independent, composed only of physical objects and physical interactions. Extreme or absolute materialism is known as *materialistic monism*, the theory that all reality is derived from one substance. Modern materialism has been largely influenced by the *theory of evolution* as described in subsequent sections.

Plato developed the school of *transcendentalism* by arguing for a higher reality (metaphysics) than that found in sense experience, and for a higher knowledge of reality (epistemology) than that achieved by human reason. Transcendentalism stems from the division of reality into a realm of spirit and a realm of matter. It affirms the existence of absolute goodness characterized as something beyond description and as knowable ultimately only through intuition. Later, religious philosophers applied this concept of transcendence to divinity, maintaining that God can be neither described nor understood in terms that are taken from human experience. This doctrine was preserved and advanced by Muslim philosophers, such as Al-Kindi (800–873), Al-Farabi (870–950), Ibn Sina (980–1037), and Ibn Rushd (1128–1198), and adopted and used by Christian philosophers, such as Aquinas (1224–1274) in the medieval period as described in subsequent sections.

Epistemology deals with issues such as the definition of knowledge and related concepts, the sources and criteria of knowledge, the kinds of knowledge possible and the degree to which each is certain, and the exact relation between the one who knows and the object known. Knowledge can be based on *priori*, knowledge derived from reason alone, and *posteriori*, knowledge gained by reference to intuitions or the facts of experience. Epistemology can be divided into *rationalism*, inquiry based on *a priori* principles — knowledge based on reason, and *empiricism*, inquiry based on *a posteriori* principles — knowledge based on experience.

Philosophical views on knowledge evolved over time. The subsequent sections summarize these views on knowledge and describe their evolution into contemporary schools. The presentation in these sections is drawn on the works of selected philosophers who either had great influence or are representatives of their respective periods. Solomon and Higgins (1996), Russell (1975), Popkin (2000), Durant (1991), and Honderich (1995) are

Table 1.2 Knowledge Views during the PreSocratics Period

Philosophers (Year)	Nature of Knowledge
Gorgias (483–378 BC)	Stated that knowledge does not exit nor can be communicated if existed.
Heraclitus (535–475 BC)	Maintained that wisdom is not the knowledge of many things; it is the clear knowledge of one thing only. Perfect knowledge is only given to the Gods, but a progress in knowledge is possible for "men."
Empedocles (c. 450 BC)	Distinguished between the world as presented to our senses (*kosmos aisthetos*) and the intellectual world (*kosmos noetos*).

recommended sources for additional details on any of the views presented in these sections.

1.2.3 Historical developments and perspectives

1.2.3.1 The preSocratic period

The *preSocratic* period includes Gorgias (483-378 BC), Heraclitus (535–475 BC), and Empedocles (c. 450 BC). Gorgias argued that nothing exists and that knowledge does not exist, nor could it be communicated to others if it existed. Heraclitus defined wisdom not as the knowledge of many things but as the clear knowledge of one thing only, and he believed in perfect knowledge given only to the Gods; however, a progress in knowledge is possible for "men." Empedocles distinguished between the world as presented to our senses (*kosmos aisthetos*) and the intellectual world (*kosmos noetos*). Table 1.2 provides a summary of these views.

1.2.3.2 The Socratic period

The *Socratic* period includes Socrates (469–399 BC), Antisthenes (440-370 BC), and Euclid (430–360 BC). The works of Socrates are available only through the descriptions of other philosophers such as Antisthenes and Euclid. Socrates' contribution to philosophy was essentially in ethics by his teaching concepts such as justice, love, virtue, and self-knowledge. He believed that all vice was the result of ignorance, and that knowledge was virtue. Socrates taught that every person has full knowledge of ultimate truth contained within the soul and needs only to be spurred to conscious reflection in order to become aware of it. Socrates employed two forms of philosophical inquiry, induction and definition. He considered dialectic thinking to be the highest method of speculative thought. Antisthenes defined happiness as a branch of knowledge that could be taught, and once acquired could not be lost. Euclid stated that knowledge is virtue. If knowledge is virtue, it can therefore be the knowledge only of the ultimate being. Table 1.3 provides a summary of these views.

Table 1.3 Knowledge Views during the Socrates Period

Philosophers (Year)	Nature of Knowledge
Antisthenes (440–370 BC)	Maintained that happiness is a branch of knowledge that could be taught, and that once acquired could not be lost.
Euclid (430–360 BC)	Maintained that knowledge is virtue. If knowledge is virtue, it can only be the knowledge of the ultimate being.

1.2.3.3 *The Plato and Aristotle period*

The *Plato and Aristotle* period includes Protagoras (485-415 BC), Plato (427–347 BC, see Figure 1.1), and Aristotle (384–322 BC, see Figure 1.2). Protagoras defined knowledge to be relative since it is based on individual experiences.

Plato's answer to Socrates' question, what makes a kind of thing the kind of thing it is, was that the *form itself* does so, and that the form is something different from the thing or object, having an eternal existence of its own. Thus, beautiful things are beautiful because they partake of beauty itself, and just acts are just insofar as they partake of justice itself, and so forth. The highest form was that of the good. Most of Plato's philosophy is concerned with metaphysics as provided in the *theory of reality*. According to this theory, reality or truth is provided by forms or ideas such as justice itself. These forms constitute the basis for reality and exist separately from the objects that are abstracted by the human senses. Humans in turn describe these objects as pale copies of the forms. Plato stated that knowledge exists based on unchanging and invisible forms or ideas. Objects that are sensed are imperfect copies of the pure forms. Genuine knowledge about these forms can be achieved only by abstract reasoning through philosophy and mathematics. Like Socrates, Plato regarded ethics as the highest branch of knowledge; he stressed the intellectual basis of virtue and identified virtue with wisdom. Plato rejected empiricism, the claim that knowledge is derived from sense experiences since propositions derived from sense experiences have, at most, a degree of belief and are not certain. Plato's *theory of forms* was intended to explain how one comes to know and also how things have come to be as they are; i.e., the theory is both an *epistemological* (theory of knowledge) and an *ontological* (theory of being) thesis.

The word *Platonism* refers both to the doctrines of Plato and to the manner or tradition of philosophizing that he founded. Often in philosophy, Platonism is virtually equivalent to idealism or intrinsicism since Plato was the first Western philosopher to claim that reality is fundamentally something ideal or abstract and that knowledge largely consists of insight into or perception of the ideal. In common usage, the adjective *Platonic* refers to the ideal; for example, Platonic love is the highest form of love that is nonsexual or nonphysical.

Plato recognized that knowledge is better than opinions. For someone to know what piety is, she or he must know it through the form, which can

Figure 1.1 Bust of Plato. (©Archivo Iconografico, S.A./CORBIS. With permission.)

only be thought and not sensed. Thus knowledge belongs to an invisible, intangible, insensible world of the intellect, while in the visible, tangible, sensible world we have only opinions. The intelligible world is more real and true than the sensible world, as well as being more distinct.

Reality, truth, and distinctness can be made for both invisible and visible worlds or realms. Within each realm, there is a further division. In the realm of the visible, there are real objects and their images, such as shadows and mirror images. These images give us the lowest grade or level of belief, mere conjectures. By seeing a shadow of an object, very little information about the specific object is gained. Similarly, there is a division within the intelligible realm, between the forms themselves and images of the forms. Knowledge of the forms themselves through reason is the highest kind of knowledge, while knowledge of the images of the forms through understanding the images is a lower form. Our opinions about the objects of the world are developed through the use of the senses, by observation. Humans can observe what things tend to go together all the time and thus develop the opinion that those things belong together. Humans might try to understand objects of the visible world by using senses, making assumptions, and exploring what follows from these interpretations and assumptions using logic. The use of assumptions can enable us to generate laws that explain why things go together the way they do. For example, Newton assumed that bodies in motion tend to stay in motion, and bodies at rest tend to stay at rest, unless some outside agency acts on them. This assumption about inertia helped him generate further principles about motion, but it is not itself proved. It can be treated as an *unexamined assumption*, in Plato's terms. This method of proceeding based on assumptions is not the best way possible for knowledge expansion since ideally it is preferred to use forms as bases for explaining other things. The forms are not only what give us knowledge, but they also can be what give things their reality. The sun casts light upon the earth, allowing us to see what is there, and it also supplies the energy through which things grow and prosper. Accordingly, the form of the good gives to the sensible world the reality it has.

The works of Plato formed the basis for *Neoplatonism*, founded by Plotinus (205–270), which greatly influenced medieval philosophers. Aristotle followed Plato as his student; however, Aristotle maintained that knowledge can be derived from sense experiences, a departure from Plato's thoughts. Knowledge can be gained either directly or indirectly by deduction using logic. For Aristotle, form and matter were inherent in all things and inseparable. Aristotle rejected the Platonic doctrine that knowledge is innate and insisted that it can be acquired only by generalization from experiences, emphasizing empiricism by stating that, "there is nothing in the intellect that was not first in the senses." Table 1.4 provides a summary of the views during this period.

Table 1.4 Knowledge Views during the Plato and Aristotle Periods

Philosophers (Year)	Nature of Knowledge
Protagoras (485–415 BC)	Maintained that knowledge is relative since it is based on individual experiences
Plato (427–347 BC)	Maintained that knowledge can exist based on unchanging and invisible *Forms* or *Ideas*. Objects that are sensed are imperfect copies of the pure forms. Genuine knowledge about these forms can be achieved only by abstract reasoning through philosophy and mathematics.
Aristotle (384–322 BC)	Followed Plato, but maintained that knowledge is derived from sense experiences. Knowledge can be gained either directly or indirectly by deduction using logic.

Figure 1.2 Portrait of Aristotle (From ©Leonard de Selva/CORBIS. With permission.)

1.2.3.4 The Hellenistic period

The *Hellenistic* period includes Epicurus (341–271 BC), Epictetus (55–135 BC), and Pyrrho (360–270 BC). Epicurus and Epictetus argued that philosophy should be a means not an end. Pyrrho argued for *skepticism* in logic and philosophy by denying the possibility of attaining any knowledge of reality apart from human perceptions. Table 1.5 provides a summary of these views.

1.2.3.5 The Medieval Period

The *Medieval* period can be characterized as an Islamic-Arabic period that resulted in translating, preserving, commenting on, and providing Europe with the works of Greek philosophers. Also, the philosophers of this period maintained and strengthened the school of rationalism and laid the foundation of *empiricism*. The philosophers of this period were influenced by Plato, Aristotle, and Plotinus who founded *Neoplatonism*, a term first used by German philosophers in the 18th century to describe a perceived development in the history of Platonism. Plotinus (205–270) is generally recognized as the founder of Neoplatonism. Plotinus' principal assumptions can be stated crudely as follows:

Table 1.5 Knowledge Views during the Hellenistic Period

Philosophers (Year)	Nature of Knowledge
Epicurus (341–271 BC) & Epictetus (55–135 CE)	Said philosophy is a means not an end.
Pyrrho (360–270 BC)	Argued for skepticism in logic and philosophy.

1. Truth exists and that it is the way the world exists in the mind or the intellect;
2. The awareness of the world as it exists in the intellect is knowledge; and
3. Two kinds of truth exist, contingent and necessary truth; for example, a contingent truth may be that ten coins are in a pocket, and a necessary truth is that four plus six equals ten.

Plotinus' innovations in Platonism were developed in his essays, the *Enneads*, which comes from the Greek word for the number nine; the essays are divided into nine groups. These groups cover ethical matters, natural philosophy, cosmology, the soul, intellect, knowledge, eternal truth, being, numbers, and the One. These innovations gave rise to Islamic Neoplatonism.

This period includes leading philosophers such as Al-Kindi (800–873), Al-Farabi (870–950), Ibn Sina (named Avicenna by the West, 980–1037), Ibn Rushd (named Averroes by the West, 1128–1198, see Figure 1.3), and Aquinas (1224–1274). Al-Kindi translated, preserved, and commented on Greek works during the Arabic civilization.

Figure 1.3 Ibn Rushd (named Averroes by the West, 1128-1198).

Al-Farabi carried the thoughts of Aristotle and was named the Second Teacher, Aristotle being the first. According to him, logic was divided into *Idea* and *Proof*. Al-Farabi made use of the logical treatises of Aristotle and employed arguments for the existence of God based upon those of Aristotle's metaphysics. The arguments were designed to provide a rational foundation for orthodox monotheism, and many of these arguments made their way into the Christian tradition later in the 13th century. Ibn Sina effectively synthesized Aristotelian, Neoplatonic, and Islamic thoughts. Ibn Rushd was named the Commentator and the Philosopher. His primary work (*Tuhafut al-Tuhafut* translated from Arabic as *The Incoherence of Limiting Rationalism*) was critical of the works of medieval philosophers in limiting rationalism and moving towards faith and revelation. For example, Al-Ghazali (1058–1128) in his work *Tuhafut al-Falasefah*, translated from Arabic as the *Incoherence of the Philosophers*, argued for less rationalism and more faith. This debate led to less rationalism in Islamic-Arabic thoughts and more of it in European thought, preparing for modern philosophy and the renaissance of Europe. Ibn Rushd attempted to overcome the contradictions between Aristotelian philosophy and revealed religion by distinguishing between two separate systems of truth: a scientific body of truths based on reason, and a religious body of truths based on revelation. This is called the *double-truth doctrine*, and influenced many Muslim, Jewish, and Christian philosophers.

Aquinas followed the schools of Plato and Aristotle and emphasized religious belief and faith. Following Neoplatonists, he considered the soul a higher form of existence than the body, and taught that knowledge results from the contemplation of Platonic ideas that have been purified of both sensation and imagery. He argued that the truths of natural science and philosophy are discovered by reasoning from facts of experiences, and the tenets of revealed religion — the doctrine of the Trinity, the creation of the world, and other articles of Christian dogma — are beyond rational comprehension, although not inconsistent with reason, and must be accepted on faith. Table 1.6 provides a summary of these views.

1.2.3.6 The Renaissance

The *Renaissance* included Bacon (1561–1626), Galileo (1564–1642), Newton (1642–1727), and Montaigne (1533–1592). Bacon denounced reliance on authority and verbal argument, criticized Aristotelian logic as useless for the discovery of new laws, and formulated rules of *inductive inference*. Galileo explained and defended the foundations of a thoroughly *empirical* view of the world and created the science of *mechanics*, which applied the principles of geometry to the motions of bodies, and that relied heavily on experimentation and empirical thoughts. Newton applied mathematics to the study of nature by formulating laws of universal gravitation and motion that explain how objects move on Earth, as well as through the heavens. Montaigne belongs to the *skepticism* school with his motto "what do I know." Table 1.7 provides a summary of these views.

Chapter one: Knowledge and ignorance 13

Table 1.6 Knowledge Views during the Medieval Period

Philosophers (Year)	Nature of Knowledge
Plotinus (205–270)	Plotinus' principal assumptions can be stated crudely as follows: (1) truth exists and that it is the way the world exists in the mind or the intellect; (2) the awareness of the world as it exists in the intellect is knowledge; and (3) two kinds of truth exist, the contingent and the necessary; for example, a contingent truth is that ten coins are in a pocket, and a necessary truth is that four plus six equals ten.
Al-Kindi (800–873)	Translated, preserved, and commented on Greek works.
Al-Farabi (870–950)	Carried the thoughts of Aristotle and was named the *Second Teacher* with Aristotle as the first. According to him logic was divided into *Idea* and *Proof*.
Ibn Sina (980–1037)	Synthesized Aristotelian, Neoplatonic, and Islamic thoughts.
Ibn Rushd (1128–1198)	Wrote a primary work (*Tuhafut al-Tuhafut*) critical of the works of medieval philosophers on limiting *rationalism* and moving to faith. Prepared for modern philosophy.
Aquinas (1224–1274)	Followed the schools of Plato and Aristotle and added religious belief and faith.

Table 1.7 Knowledge Views during the Renaissance

Philosophers (Year)	Nature of Knowledge
Bacon (1561–1626)	Criticized Aristotelian logic as useless for the discovery of new laws; and formulated rules of *inductive inference*.
Galileo (1564–1642)	Explained and defended the foundations of a thoroughly *empirical* view of the world by creating the science of *mechanics*, which applied the principles of geometry to the motions of bodies.
Newton (1642–1727)	Applied mathematics to the study of nature.
Montaigne (1533–1592).	Belongs to the skepticism school with his motto "what do I know."

1.2.3.7   *The 17th century*

The *17th century* includes Descartes (1596–1650), Spinoza (1632–1677), and Locke (1632–1704). Descartes, the father of modern philosophy, identified *rationalism*, which is sometimes called *Cartesian rationalism*. Rationalism is a system of thought that emphasizes the role of reason and *priori* principles in obtaining knowledge; Descartes used the expression "I think, therefore I am." He also believed in the dualism of mind (thinking substance) and body (extended substance). Spinoza termed metaphysical (i.e., cosmological) concepts such as *substance* and *mode*, *thought* and *extension*, *causation* and *parallelism*, and *essence* and *existence* to reconcile concepts related to God,

Table 1.8 Knowledge Views during the 17th Century

Philosophers (Year)	Nature of Knowledge
Descartes (1596–1650)	As the father of modern philosophy, identified *rationalism* as a system of thought that emphasized the role of reason and *priori* principles in obtaining knowledge. He also believed in the dualism of mind (thinking substance) and body (extended substance).
Spinoza (1632–1677)	Termed metaphysical (i.e., cosmological) concepts such as substance and mode, thought and extension, causation and parallelism, and essence and existence.
Locke (1632–1704)	Identified *empiricism* as a doctrine that affirms that all knowledge is based on experience, especially sense perceptions, and on *posteriori* principles. Locke believed that human knowledge of external objects is always subject to the errors of the senses, and concluded that one cannot have absolutely certain knowledge of the physical world.

substance, and nature. Locke identified *empiricism* as a doctrine affirming that all knowledge is based on experience, especially sense perceptions, and on *posteriori* principles. Empiricism denies the possibility of spontaneous ideas or a priori thought. Locke distinguished two sources of experience: sensation-based knowledge of the external world and reflection-based knowledge of the mind. Locke believed that human knowledge of external objects is always subject to the errors of the senses and concluded that one cannot have absolutely certain knowledge of the physical world. Table 1.8 provides a summary of these views.

1.2.3.8 The 18th century

The *18th century* includes leading philosophers such as Berkeley (1685–1753), Hume (1711–1776), and Kant (1724–1804). Berkeley is the founder of the school of *idealism*. He agreed with Locke that knowledge comes through ideas, i.e., sensation of the mind, but he denied Locke's belief that a distinction can be made between ideas and objects. Berkeley held that matter cannot be conceived to exist independent of the mind, and that the phenomena of sense experiences can be explained only by supposing a deity that continually evokes perception in the human mind. Extending Locke's doubts about knowledge of the world outside the mind, Berkeley argued that no evidence exists for the existence of such a world because the only things that one can observe are one's own sensations, and these are in the mind. Berkeley established the epistemological view *phenomenalism*, a theory of perception suggesting that matter can be analyzed in terms of sensations, preparing the way for the *positivist* movement in modern thought. Hume asserted that all metaphysical things that cannot be directly perceived are meaningless. He divided all knowledge into two kinds: (1) *relations* of *ideas*, i.e., the knowledge found in mathematics and logic, which is exact and certain but provides no

Table 1.9 Knowledge Views during the 18th Century

Philosophers (Year)	Nature of Knowledge
Berkeley (1685–1753)	Agreed with Locke that knowledge comes through *ideas*, i.e., sensation of the mind, but he denied Locke's belief that a distinction can be made between ideas and objects.
Hume (1711–1776)	Asserted that all metaphysical things that cannot be directly perceived are meaningless. Hume divided all knowledge into two kinds: relations of *ideas*, i.e., the knowledge found in mathematics and logic which is exact and certain but provides no information about the world, and *matters of fact*, i.e., the knowledge derived from sense perceptions. Furthermore, he held that even the most reliable laws of science might not always remain true.
Kant (1724–1804)	Provided a compromise between empiricism and rationalism by combining both types, and distinguished three knowledge types: (1) an *analytical priori*, (2) a *synthetic posteriori*, and (3) a *synthetic priori*.

information about the world; and (2) *matters of fact*, i.e., the knowledge derived from sense perceptions. Furthermore, he held that even the most reliable laws of science might not always remain true. Kant provided a compromise between empiricism and rationalism by combining them. He distinguished between three types of knowledge: (1) an *analytical priori*, which is exact and certain, but also uninformative because it makes clear only what is contained in definitions; (2) a *synthetic posteriori*, which conveys information about the world learned from experience but is subject to the errors of the senses; and (3) a *synthetic priori*, which is discovered by pure intuition and is both exact and certain, for it expresses the necessary conditions that the mind imposes on all objects of experience. The 19th century philosophers argued over the existence of the above third type of knowledge. Table 1.9 provides a summary of these views.

1.2.3.9 *The 19th century*

The *19th century* includes leading philosophers such as Hegel (1770–1831), Comte (1798–1857), Marx (1818–1883) and Engels (1820–1895), and Nietzsche (1844–1900). Hegel claimed, as a *rationalist*, that absolutely certain knowledge of reality can be obtained by equating the processes of thought, of nature, and of history. His *absolute idealism* was based on a dialectical process of thesis, antithesis, and synthesis as cyclical and ongoing process; where a thesis is any idea or a historical movement, an antithesis is a conflicting idea or movement, and synthesis overcomes the conflict by reconciling a higher level of truth contained in both. Therefore, conflict and contradiction are regarded as necessary elements of truth, and truth is regarded as a process rather than a fixed state of things. He considered the *Absolute Spirit* to be the sum of all reality with reason as a master of the world, i.e., by stating that

Table 1.10 Knowledge Views during the 19th Century

Philosophers (Year)	Nature of Knowledge
Hegel (1770–1831)	Claimed as a rationalist that absolutely certain knowledge of reality can be obtained by equating the processes of thought, of nature, and of history. His absolute idealism was based on a dialectical process of thesis, antithesis, and synthesis as cyclical and ongoing process.
Comte (1798–1857)	Brought attention to the importance of sociology as a branch of knowledge and extended the principles of positivism, the notion that empirical sciences are the only adequate source of knowledge.
Marx (1818–1883) and Engels (1820–1895)	Developed the philosophy of *dialectical materialism*, based on the logic of Hegel.
Nietzsche (1844–1900)	Concluded that traditional philosophy and religion are both erroneous and harmful, and traditional values (represented primarily by Christianity) had lost their power in the lives of individuals. Therefore, there are no rules for human life, no absolute values, and no certainties on which to rely.

"what is rational is real and what real is rational." Comte brought attention to the importance of sociology as a branch of knowledge and extended the principles of *positivism*, the notion that empirical sciences are the only adequate source of knowledge. Marx and Engels developed the philosophy of *dialectical materialism*, based on the logic of Hegel, leading to *social Darwinism*, based on the *theory of evolution* developed by the British naturalist Charles Darwin. According to social Darwinism, living systems compete in a struggle for existence in which natural selection results in "survival of the fittest." Marx and Engels derived from Hegel the belief that history unfolds according to *dialectical laws* and that social institutions are more concretely real than physical nature or the individual mind. Nietzsche concluded that traditional philosophy and religion are both erroneous and harmful, and that traditional values (represented primarily by Christianity) had lost their power in the lives of individuals. He concluded that there are no rules for human life, no absolute values, and no certainties on which to rely. Table 1.10 provides a summary of these views.

1.2.3.10 The 20th century

The *20th century* includes leading philosophers such as Bradley (1846–1924), Royce (1855–1916), Peirce (1839–1914), Dewey (1859–1952), Husserl (1859–1938), Russell (1872–1970), Wittgenstein (1889–1951), and Austin (1911–1960). Bradley maintained that reality was a product of the mind rather than an object perceived by the senses. Like Hegel, he also maintained that nothing is altogether real except the *Absolute*, the totality of everything which transcends contradiction. Everything else, such as religion, science, moral precept, and even common sense, is contradictory. Royce believed in an *absolute truth*, and held that human thought and the external world were unified. Peirce developed *pragmatism* as a theory of meaning, in particular the meaning of concepts used in science. The only rational way to increase knowledge was to form mental habits that would test ideas through observation and experimentation leading to an evolutionary process of knowledge for humanity and society, i.e., a *perpetual state of progress*. He believed that the truth of an idea or object could be measured only by empirical investigation of its usefulness. Pragmatists regarded all theories and institutions as tentative hypotheses and solutions, and they believed that efforts to improve society must be geared toward problem solving in an ongoing process of progress. Pragmatism sought a middle ground between traditional metaphysical ideas about the nature of reality and the radical theories of *nihilism* and *irrationalism* which had become popular in Europe at that time. They did not believe that a single absolute idea of goodness or justice existed, but rather that these concepts were relative and depended on the context in which they were being discussed.

Pierce influenced a group of philosophers, called *logical positivists*, who emphasized the importance of scientific verification, and rejected personal experience as the basis of true knowledge. Dewey further developed pragmatism into a comprehensive system of thought that he called *experimental naturalism*, or *instrumentalism*. *Naturalism* regards human experience, intelligence, and social communities as ever-evolving mechanisms; therefore human beings could solve social problems using their experience and intelligence and through inquiry. He considered traditional ideas about knowledge and absolute reality or absolute truth to be incompatible with a Darwinian world view of progress; therefore, they must be discarded or revised.

Husserl developed *phenomenology* as an elaborate procedure by which one is said to be able to distinguish between the way things appear to be and the way one thinks they really are. Russell revived empiricism and expanded to *epistemology* as a field. He attempted to explain all factual knowledge as constructed out of immediate experiences. Wittgenstein developed *logical positivism* that maintained (1) only scientific knowledge exists, (2) any valid knowledge must be verifiable in experience, and (3) a lot of previous philosophy was neither true nor false, but literally meaningless. In his words, "philosophy is a battle against the bewitchment of our intelligence by means of language." He viewed philosophy as a linguistic analysis and "language games," leading to his work *Tractatus Logico-Philosophicus* (1921) that asserted language is composed of complex propositions that can be analyzed into less complex propositions until one arrives at simple or elementary propositions. This view of decomposing complex language propositions has a parallel in our view of the world to be composed of complex facts that can be analyzed into less complex facts until one arrives at simple "picture atomic facts or states of affairs." Wittgenstein's *picture theory of meaning* required and built on atomic facts pictured by the elementary propositions. Therefore, only propositions that picture facts are the propositions of science that can be considered cognitively meaningful. Metaphysical, ethical, and theological statements, on the other hand, are not meaningful assertions. Wittgenstein's work influenced the work of Russell in developing the theory of *logical atomism*.

Russell, Wittgenstein, and others formed the core of the Vienna Circle that developed *logical positivism* in which philosophy is defined by its role in clarification of meaning, not the discovery of new facts or the construction of traditional metaphysics. They introduced strict principles of verifiability to reject as meaningless the nonempirical statements of metaphysics, theology, and ethics, and they regarded as meaningful only statements reporting empirical observations, taken together with the tautologies of logic and mathematics. Austin developed the *speech-act* theory, which states that many utterances do not merely describe reality, but also have an effect on reality, insofar as they too are the performance of some act. Table 1.11 provides a summary of these views.

1.2.4 Knowledge, information, and opinions

Many disciplines of engineering and the sciences rely on the development and use of predictive models that in turn require knowledge and information and sometimes subjective opinions of experts. Working definitions for knowledge, information, and opinions are needed for this purpose. In this section, these definitions are provided with some limitations and discussions of their uses.

Knowledge can be based on *evolutionary epistemology* using an *evolutionary model*. It can be viewed to consist of two types, *nonpropositional* and *propositional* knowledge. Nonpropositional knowledge can be further broken down into *know-how* and *concept knowledge*, and *familiarity knowledge* (commonly called *object knowledge*). Know-how and concept knowledge require someone to know how to do a specific activity, function, procedure, etc., such as riding a bicycle. Concept knowledge can be empirical in nature. In evolutionary epistemology, know-how knowledge is viewed as a historical antecedent to propositional knowledge. Object knowledge is based on a direct acquaintance with a person, place, or thing; for example, Mr. Smith knows the President of the United States. Propositional knowledge is based on propositions that can be either true or false; for example, Mr. Smith knows that the Rockies are in North America (Sober 1991, and di Carlo 1998). This proposition can be expressed as

$$\text{Mr. Smith knows that the Rockies are in North America} \quad (1\text{-}1a)$$

$$S \text{ knows } P \quad (1\text{-}1b)$$

where S is the subject, i.e., Mr. Smith, and P is the claim "the Rockies are in North America." Epistemologists require the following three conditions for making this claim in order to have a true proposition:

Table 1.11 Knowledge Views during the 20th Century

Philosophers (Year)	Nature of Knowledge
Bradley (1846–1924)	Maintained that reality was a product of the mind rather than an object perceived by the senses; like Hegel, nothing is altogether real except the Absolute, the totality of everything which transcends contradiction. Everything else, such as religion, science, moral precept, and even common sense, is contradictory.
Royce (1855–1916)	Believed in an absolute truth and held that human thought and the external world were unified.
Peirce (1839–1914)	Developed *pragmatism* as a theory of meaning, in particular, the meaning of concepts used in science. The only rational way to increase knowledge was to form mental habits that would test ideas through observation and experimentation leading to an evolutionary process for humanity and society, i.e., a perpetual state of progress. He believed that the truth of an idea or object could only be measured by empirical investigation of its usefulness.
Dewey (1859–1952)	Further developed pragmatism into a comprehensive system of thought that he called *experimental naturalism*, or *instrumentalism*. Naturalism regards human experience, intelligence, and social communities as ever-evolving mechanisms; therefore human beings could solve social problems using their experience and intelligence, and through inquiry.
Husserl (1859–1938)	Developed *phenomenology* as an elaborate procedure by which one is said to be able to distinguish between the way things appear to be, and the way one thinks they really are.
Russell (1872–1970)	Revived empiricism and expanded it to *epistemology* as a field.
Wittgenstein (1889–1951)	Developed *logical positivism* that maintained that only scientific knowledge exists verifiable by experience. He viewed philosophy as a linguistic analysis and "language games" leading to his work *Tractatus Logico-Philosophicus* (1921) that asserted language, or the world, are composed of complex propositions, or facts, that can be analyzed into less complex propositions arriving at elementary propositions, or into less complex facts, arriving at simple "picture atomic facts or states of affairs," respectively.
Austin (1911–1960)	Developed the *speech-act theory*, in which language utterances might not describe reality and can have an effect on reality.

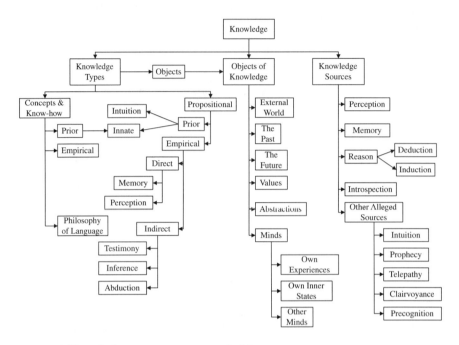

Figure 1.4 Knowledge types, sources, and objects.

- S must believe P,
- P must be true, and
- S must have a reason to believe P; i.e., S must be justified in believing P.

The justification in the third condition can take various forms; however, simplistically it can be taken as justification through rational reasoning or empirical evidence. Therefore, propositional knowledge is defined as a body of propositions that meet the conditions of *justified true belief* (JTB). This general definition does not satisfy a class of examples, the *Gettier problem*, initially revealed in 1963 by Edmund Gettier (Austin, 1998) as provided in Example 1.1. Gettier showed that we can have highly reliable evidence and still not have knowledge. Also, someone can skeptically argue that as long as it is possible for S to be mistaken in believing P (i.e., not meet the third condition), the proposition is false. This argument, sometimes called a Cartesian argument, undermines empirical knowledge. In evolutionary epistemology, this high level of scrutiny is not needed, and it need not be satisfied in engineering and the sciences. According to evolutionary epistemology, true beliefs can be justified causally from reliably attained law-governed procedures, where law refers to a natural law. Sober (1991) noted that there are very few instances, if ever, where we have perfectly infallible evidence. Almost all of our common sense beliefs are based on evidence that is not infallible even though some may have overwhelming reliability. The presence of a small doubt in meeting the justification condition does not make our evidence infallible but only reliable. Evidence reliability and infallibility arguments form the basis of the *reliability theory of knowledge*. Figure 1.4 shows a breakdown of knowledge by types, sources, and objects that was based on a summary provided by Honderich (1995).

In engineering and the sciences, knowledge can be defined as a body of JTB, such as, laws, models, objects, concepts, know-how, processes, and principles, acquired by humans about a system of interest, where the justification condition can be met based on the reliability theory of knowledge. The most basic knowledge category is cognitive knowledge (*episteme*) that can be acquired by human senses. The next level is based on correct reasoning from hypotheses such as mathematics (*dianoi*). The third category moves us from intellectual categories to categories that are based on the realm of appearances and deception and are based on propositions. This third category is belief (*pistis* — the Greek word for faith, denoting intellectual and/or emotional acceptance of a proposition). It is followed by conjecture (*eikasia*) in which knowledge is based on inference, theorization, or prediction based on incomplete or reliable evidences. The four categories are shown in Figure 1.5 and also define the knowledge box in Figure 1.6. These categories constitute the human cognition of human knowledge that might be different from a future state of knowledge achieved by an evolutionary process, as shown in Figure 1.6. The *pistis* and *eikasia* categories are based on expert judgment and opinions regarding system issues of interest. Although the *pistis* and *eikasia* knowledge categories might by marred with uncertainty, they are a certainty sought after in many engineering disciplines and the sciences, especially by decision and policy makers.

Information can be defined as sensed objects, things, places, processes, and information and knowledge communicated by language and multimedia. Information can be viewed as a preprocessed input to our intellect system of cognition, and as knowledge acquisition and creation. It can lead to knowledge through investigation, study, and reflection. However, knowledge and information about the system might not constitute the eventual evolutionary knowledge state about the system as a result of not meeting the justification condition in JTB or the ongoing evolutionary process or both. Knowledge is defined in the context of the humankind, evolution, language and communication methods, and social and economic dialectic processes; it cannot be removed from them. As a result, knowledge would always reflect the imperfect and evolutionary nature of humans, which is attributed to their reliance on their senses for information acquisition; their dialectic processes; and their mind for extrapolation, creativity, reflection, and imagination, with associated biases as a result of preconceived notions due to time asymmetry, specialization, and other factors. An important dimension in defining the state of knowledge and truth about a system is nonknowledge or ignorance.

Opinions rendered by experts, that are based on information and existing knowledge, can be defined as preliminary propositions with claims that are not fully justified or that are justified with adequate reliability but are not necessarily infallible. Expert opinions are seeds of propositional knowledge that do not meet one or more of the conditions required by the JTB and the reliability theory of knowledge. They are valuable as they might lead to knowledge expansion, but decisions based on them sometimes might be risky propositions, since their preliminary nature means they might be proven false by others or in the future.

The relationships among knowledge, information, opinions, and evolutionary epistemology are schematically shown in Figure 1.5. The dialectic processes include communication methods such as languages, visual and audio formats, and other forms. Also, they include economic, class, schools of thought, political and social dialectic processes within peers, groups, colonies, societies, and the world.

Example 1.1. A Gettier problem: a Ferrari owner or not

Knowledge was defined as accumulations of justified true beliefs (JTB). This general definition does not satisfy any of a class of examples known collectively as the Gettier problem, initially revealed in 1963 by Edmund Gettier (Austin, 1998). For example, a teacher has two students, Mr. Nothavit and Mr. Havit, in her class. Mr. Nothavit seems to be the proud owner of an expensive car (a Ferrari). Mr. Nothavit says he owns one, drives one around, and has papers that state that he owns the car, but he does not actually own a Ferrari. On the basis of this evidence, the teacher concludes that the proposition "someone in her class owns a Ferrari" is true. This proposition is true enough, but only because Mr. Havit, who shows no signs of Ferrari ownership, secretly owns one. Therefore, it seems that the three conditions (truth, justification, and belief) of knowledge have been met, but that there is no knowledge. So it is true that a Ferrari owner is a member of her class, the teacher accepts that it is true, and she is completely justified in so accepting that he is.

1.3 *Cognition and cognitive science*

Cognition can be defined as mental processes of receiving and processing information for knowledge creation and behavioral actions. Cognitive science is the interdisciplinary study of mind and intelligence (Stillings, 1995). Cognitive science deals with philosophy, psychology, artificial intelligence, neuroscience, linguistics, and anthropology. The intellectual origins of cognitive science started in the mid-1950s when researchers in several fields used complex representations and computational procedures to develop theories on how the mind works.

The origin of cognitive science can be taken as the theory of knowledge and the theory of reality of the ancient Greeks, when philosophers such as Plato and Aristotle tried to explain the nature of human knowledge. The study of mind remained the province of philosophy until the nineteenth century, when experimental psychology was developed by Wilhelm Wundt

Chapter one: Knowledge and ignorance

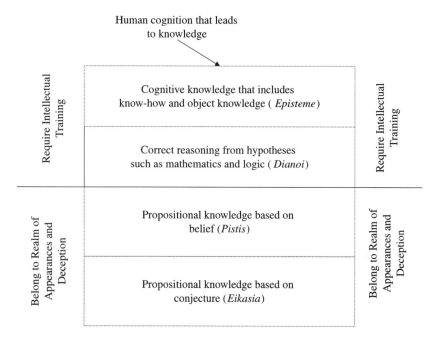

Figure 1.5 Knowledge categories and sources.

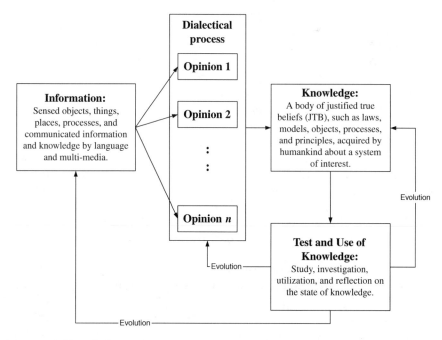

Figure 1.6 Knowledge, information, opinions, and evolutionary epistemology.

and his students by initiating laboratory methods for studying systematically mental operations. A few decades later, experimental psychology became dominated by behaviorism by which the existence of the mind was virtually denied. Behaviorists, such as J. B. Watson, argued that psychology should restrict itself to examining the relation among observable stimuli and observable behavioral responses and should not deal with consciousness and mental representations. The intellectual landscape began to change dramatically in 1956, when George Miller summarized numerous studies showing that the capacity of human thinking is limited, with short-term memory, for example, limited to around seven items. He proposed that humans compensate for memory limitations through their ability to recode information into chunks and mental representations that require mental procedures for encoding and decoding the information. Although at this time primitive computers had been around for only a few years, pioneers such as John McCarthy, Marvin Minsky, Allen Newell, and Herbert Simon were founding the field of artificial intelligence. Moreover, Noam Chomsky rejected behaviorist assumptions about language as a learned habit and proposed instead to explain language comprehension in terms of mental grammars consisting of rules.

Cognitive science is based on a central hypothesis that thinking can best be understood in terms of representational structures in the mind and computational procedures that operate on those structures (Johnson-Laird, 1988). The nature of the representations and computations that constitute thinking are not fully understood. The central hypothesis is general enough to encompass the current range of thinking in cognitive science, including connectionist theories which model thinking using artificial neural networks. This hypothesis assumes that the mind has mental representations analogous to computer data structures, and computational procedures similar to computational algorithms. The mind is considered to contain such mental representations as logical propositions, rules, concepts, images, and analogies. It uses mental procedures such as deduction, search, matching, rotating, and retrieval for interpretation, generation of knowledge, and decision making. The dominant mind-computer analogy in cognitive science has taken on a novel twist from the use of another analog — the brain. Cognitive science then works with a complex 3-way analogy among the mind, brain, and computers. Connectionists have proposed a brain-like structure that uses neurons and their connections as inspirations for data structures, and neuron firing and spreading activation as inspirations for algorithms. There is not a single computational model for the mind, since different kinds of computers and programming approaches suggest different ways in which the mind might work, ranging from serial processors, such as the commonly used computers that perform one instruction at a time, to parallel processors, such as some recently developed computers that are capable of doing many operations at once.

Cognitive science claims that the human mind works by representation and computation using empirical conjecture. Although the computational-representational approach to cognitive science has been successful in explaining many aspects of human problem solving, learning, and language use, some philosophical critics argue that it is fundamentally flawed because of the following limitations (Thagard, 1996, and Von Eckardt, 1993):

- Emotions: Cognitive science neglects the important role of emotions in human thinking.
- Consciousness: Cognitive science ignores the importance of consciousness in human thinking.
- Physical environments: Cognitive science disregards the significant role of physical environments on human thinking.
- Social factors: Human thought is inherently social and has to deal with various dialectic processes in ways that cognitive science ignores.
- Dynamical nature: The mind is a dynamical system, not a computational system.
- Quantum nature: Researchers argue that human thinking cannot be computational in the standard sense, so the brain must operate differently, perhaps as a quantum computer.

These open issues need to be considered by scientists and philosophers in developing new cognitive theories and a better understanding of how the human mind works.

1.4 Time and its asymmetry

Time and its asymmetry are crucial in defining knowledge and ignorance (Horwich, 1987). Kurt Gödel speculated in a theory, which is consistent with the *General Theory of Relativity*, that time flows from the past to the future, passing through the present, and allows for "time travel" to the past. However, we can reliably claim, based on our current technology and knowledge base, that time as a phenomenon has a unidirectional flow. Time can be viewed as a one-dimensional continuum of instants with temporally occurring events. The present (or now) is a gliding index that moves in a unidirectional form from the past to the future. "It is as if we were floating on a river, carried by the current, past the manifold events which are spread out timelessly on the bank," said Plato.

Engineers and scientists are practitioners and specialists who often try to make statements about the future. However, Aristotle asserted that contingent statements about the future have no truth value, unlike statements made about the past and present which are determinably either true or false. Events of interest can be viewed to progress in time tree-like, with fixed branches of the past and forming branches of the present. However, the future branches manifold undetermined possibilities. Many scientific and engineering laws and principles display temporal irreversibility, such as thermodynamic system changes and chemical reactions. Therefore, they can be viewed as time-asymmetric system changes and transitions. In addition, there are many physical processes that do not have temporal inverses.

Knowledge is primarily the product of the past as we know more about the past than the future. For example, we can precisely describe past daily temperatures but cannot accurately forecast future temperatures. Time asymmetry of knowledge can be attributed to several factors, of which the significant ones are

- Our limited capacity to free ourselves from the past in order to forecast in the future.
- Our inability to go back in time and verify historical claims; therefore, we are overconfident in the superiority of our present knowledge.
- The unidirectional nature of causation to the past but not the future. We tend to explain phenomena based on antecedents rather than consequences. Therefore, we assume that causes precede effects. Although, the order can be switched for some systems, as someone might be creating the effects needed for some causes. The unidirectional temporal nature of explanation might not be true all the time and sometimes can be non-verifiable. In economics, for example, incentives (i.e., consequences) can create causes and means.

Engineers and scientists tend to be preoccupied more with what will happen than what has happened. This preoccupation might result in bias and time asymmetry. Physical, social, or economic systems can be characterized by their goals as well as by their causes that entail the future, the past, and the present.

1.5 Defining ignorance in the context of knowledge

1.5.1 Human knowledge and ignorance

Generally, engineers and scientists, and even almost most humans, tend to focus on what is known and not on the unknowns. Even the English language lends itself for this emphasis. For example, we can easily state that Expert A informed Expert B, whereas we cannot *directly* state the contrary. We can only state it by using the negation of the earlier statement as "Expert A did not inform Expert B." Statements such as "Expert A misinformed Expert B," or "Expert A ignored Expert B" do not convey the same (intended) meaning. Another example is "John knows David," for which a meaningful *direct* contrary statement does not exist. The emphasis on knowledge and not on ignorance can also be noted in sociology by its having a field of study called the *sociology of knowledge* and not having *sociology of ignorance*, although Weinstein and Weinstein (1978) introduced the *sociology of non-knowledge*, and Smithson (1985) introduced the *theory of ignorance*.

Engineers and scientists tend to emphasize knowledge and information, and sometimes they intentionally or unintentionally brush aside ignorance. In addition, information (or knowledge) can be misleading in some situations because it does not have the truth content that was assigned to it, leading

Chapter one: Knowledge and ignorance 27

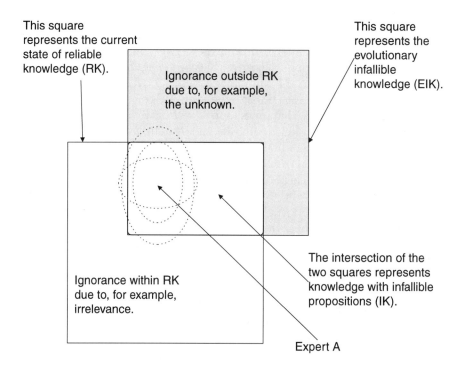

Figure 1.7 Human knowledge and ignorance.

potentially to overconfidence. In general, knowledge and ignorance can be classified as shown in Figure 1.7, using squares with crisp boundaries for the purpose of illustration. The shapes and boundaries can be made multi-dimensional, irregular, and/or fuzzy. The evolutionary infallible knowledge (EIK) about a system is shown as the top-right square in the figure and can be intrinsically unattainable due to the fallacy of humans and the evolutionary nature of knowledge. The state of reliable knowledge (RK) is shown using another square, i.e., the bottom left square, for illustration purpose. The reliable knowledge represents the present state of knowledge in an evolutionary process, i.e., a snapshot of knowledge as a set of know-how, object, and prepositions that meet justifiable true beliefs within reasonable reliability levels. At any stage of human knowledge development, this knowledge base about the system is a mixture of truth and fallacy. The intersection of EIK and RK represents the knowledge base with the infallible knowledge components (i.e., know-how, objects, and propositions). Therefore, the following relationship can be stated using the notations of set theory:

$$\text{Infallible Knowledge (IK)} = \text{EIK} \cap \text{RK} \quad (1.2)$$

where ∩ means intersection. Infallible knowledge is defined as knowledge that can survive the dialectic processes of humans and societies and that passes the test of time and use. This infallible knowledge can be schematically defined by the intersection of these two squares of EIK and RK. Based on this representation, two primary types of ignorance can be identified: (1) ignorance within the knowledge base RK due to factors such as irrelevance, and (2) ignorance outside the knowledge base due to unknown objects, interactions, laws, dynamics, and know-how.

Expert A of some knowledge about the system can be represented as shown in Figure 1.7, using ellipses for illustrative purposes. Three types of ellipses can be identified: (1) a subset of the evolutionary infallible knowledge (EIK) that the expert has learned, captured, and/or created, (2) self-perceived knowledge by the expert, and (3) perception by others of the expert's knowledge. The EIK of the expert might be smaller than the self-perceived knowledge by the expert, and the difference between the two types is a measure of overconfidence that can be partially related to the expert's ego. Ideally, the three ellipses should be the same, but commonly they are not. They are greatly affected by communication skills of experts and their successes in dialectic processes that with time might lead to evolutionary knowledge marginal advances or quantum leaps. Also, their relative sizes and positions within the infallible knowledge (IK) base are unknown. It can be noted from Figure 1.7 that the expert's knowledge can extend beyond the reliable knowledge base into the EIK area as a result of creativity and imagination by the expert. Therefore, the intersection of the expert's knowledge with the ignorance space outside the knowledge base can be viewed as a measure of creativity and imagination. Another expert (i.e., Expert B) would have her/his own ellipses that might overlap with the ellipses of Expert A and might overlap with other regions by varying magnitudes.

1.5.2 Classifying Ignorance

The state of ignorance for a person or society can be unintentional or deliberate due to an erroneous cognition state and not knowing relevant information, or ignoring information and deliberate inattention to something for various reasons, such as limited resources or cultural opposition, respectively. The latter type is a state of *conscious ignorance* which is not intentional, and once it is recognized evolutionary species try to correct that state for survival reasons, with varying levels of success. The former ignorance type belongs to the *blind ignorance* category. Therefore, ignoring means that someone can either *unconsciously* or *deliberately* refuse to acknowledge or regard, or leave out an account or consideration for relevant information (di Carlo, 1998). These two states should be treated in developing a hierarchal breakdown of ignorance.

Using the concepts and definitions from evolutionary knowledge and epistemology, ignorance can be classified based on the three knowledge sources as follows:

- *Know-how ignorance* can be related to the lack of know-how knowledge or the having of erroneous know-how knowledge. Know-how knowledge requires someone to know how to do a specific activity, function, procedure, etc., such as riding a bicycle.
- *Object ignorance* can be related to the lack of object knowledge or the having of erroneous object knowledge. Object knowledge is based on a direct acquaintance with a person, place or thing; for example, Mr. Smith knows the President of the United States.
- *Propositional ignorance* can be related to the lack of propositional knowledge or the having of erroneous propositional knowledge. Propositional knowledge is based on propositions that can be either true or false; for example, Mr. Smith knows that the Rockies are in North America.

The above three ignorance types can be cross-classified against two possible states for a knowledge agent, such as a person, of knowing their state of ignorance. These two states are

- *Nonreflective* (or *blind*) *state*: The person does not know of self-ignorance, a case of ignorance of ignorance.
- *Reflective state*: The person knows and recognizes self-ignorance. Smithson (1985) termed this type of ignorance conscious ignorance, and the blind ignorance was termed *meta-ignorance*. As a result, in some cases the person might formulate a proposition but still be ignorant of the existence of a proof or disproof. A knowledge agent's response to reflective ignorance can be either passive acceptance or a guided attempt to remedy one's ignorance that can lead to four possible outcomes: (1) a successful remedy that is recognized by the knowledge agent to be a success, leading to fulfillment; (2) a successful remedy that is not recognized by the knowledge agent to be a success, leading to searching for a new remedy; (3) a failed remedy that is recognized by the knowledge agent to be a failure, leading to searching for a new remedy; and (4) a failed remedy that is recognized by the knowledge agent to be a success, leading to blind ignorance, such as *ignoratio elenchi*, i.e., ignorance of refutation or missing the point or irrelevant conclusion.

The cross classification of ignorance is shown in Figure 1.8 in two possible forms that can be used interchangeably. Although the blind state does not feed directly into the evolutionary process for knowledge, it represents a becoming knowledge reserve. The reflective state has a survival value to evolutionary species; otherwise it can be argued that it never would have flourished (Campbell, 1974). Ignorance emerges as a lack of knowledge relative to a particular perspective from which such gaps emerge. Accordingly, the accumulation of beliefs and the emergence of ignorance constitute a dynamic process resulting in old ideas perishing and new ones flourishing

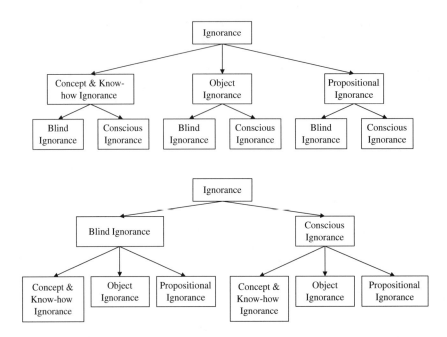

Figure 1.8 Classifying ignorance.

(Bouissac, 1992). According to Bouissac (1992), the process of scientific discovery can be metaphorically described as not only a cumulative sum (positivism) of beliefs, but also an activity geared towards relentless construction of ignorance (negativism), producing an architecture of holes, gaps, and lacunae, so to speak.

Hallden (1986) examined the concept of evolutionary ignorance in decision theoretic terms. He introduced the notion of gambling to deal with blind ignorance or lack of knowledge according to which there are times when, in lacking knowledge, gambles must be taken. Sometimes gambles pay off with success, i.e., continued survival, and sometimes they do not, leading to sickness or death.

According to evolutionary epistemology, ignorance has factitious, i.e., human-made, perspectives. Smithson (1988) provided a working definition of ignorance based on "Expert A is ignorant from B's viewpoint if A fails to agree with or show awareness of ideas that B defines as actually or potentially valid." This definition allows for self-attributed ignorance, and either Expert A or B can be attributer or perpetrator of ignorance. Our ignorance and claimed knowledge depend on our current historical setting, which is relative to various natural and cultural factors, such as language, logical systems, technologies, and standards, which have developed and evolved over time. Therefore, humans evolved from blind ignorance through gambles to a state of incomplete knowledge with reflective ignorance recognized

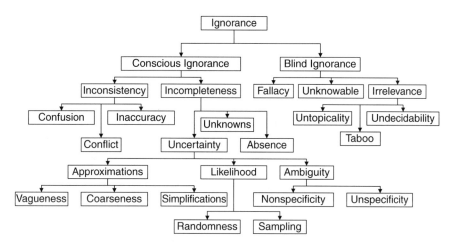

Figure 1.9 Ignorance hierarchy.

through factitious perspectives. In many scientific fields, the level of reflective ignorance becomes larger as the level of knowledge increases. Duncan and Weston-Smith (1997) stated in the *Encyclopedia of Ignorance* that, compared to our bond of knowledge, our ignorance remains atlantic. They invited scientists to state what they would like to know in their respective fields and noted that the more eminent the scientists were, the more readily and generously they described their ignorance. Clearly, before solving a problem, it needs to be articulated.

1.5.3 Ignorance hierarchy

Figures 1.6 and 1.7 express knowledge and ignorance in evolutionary terms as they are socially or factitiously constructed and negotiated. Ignorance can be viewed to have a hierarchal classification based on its sources and nature as shown in Figure 1.9 with the brief definitions provided in Table 1.12. As stated earlier, ignorance can be classified into two types, blind ignorance (also called meta-ignorance) and conscious ignorance (also called reflective ignorance).

Blind ignorance includes *not knowing* relevant know-how, objects-related information, and relevant propositions that can be justified. The unknowable knowledge can be defined as knowledge that cannot be attained by humans based on current evolutionary progressions, cannot be attained at all due to human limitations, or can be attained only through quantum leaps by humans. Blind ignorance also includes irrelevant knowledge that can be of two types: (1) relevant knowledge that is dismissed as irrelevant or ignored, and (2) irrelevant knowledge that is believed to be relevant through nonreliable or weak justification or as a result of *ignoratio elenchi*. The irrelevance type can be due to untopicality, taboo, and undecidability. Untopicality can

Table 1.12 Taxonomy of Ignorance

Term	Meaning
1. Blind ignorance	Ignorance of self-ignorance also called meta-ignorance.
1.1. Unknowable	Knowledge that cannot be attained by humans based on current evolutionary progressions, cannot be attained at all due to human limitations, or can be attained only through quantum leaps by humans.
1.2. Irrelevance	Ignoring something.
1.2.1. Untopicality	Intuitions of experts that could not be negotiated with others in terms of cognitive relevance.
1.2.2. Taboo	Socially reinforced irrelevance. Issues that people must not know, deal with, inquire about, or investigate.
1.2.3. Undecidability	Issues that cannot be designated true or false because they are considered insoluble; or solutions that are not verifiable; or *ignoratio elenchi*.
1.3. Fallacy	Erroneous belief due to misleading notions.
2. Conscious ignorance	A recognized self-ignorance through reflection.
2.1. Inconsistency	Inconsistency in knowledge can be attributed to distorted information as a result of inaccuracy, conflict, contradiction, and/or confusion.
2.1.1. Confusion	Wrongful substitutions.
2.1.2. Conflict	Conflicting or contradictory assignments or substitutions.
2.1.3. Inaccuracy	Bias and distortion in degree.
2.2. Incompleteness	Incomplete knowledge due to absence or uncertainty.
2.2.1. Absence	Incompleteness in kind.
2.2.2. Unknowns	The difference between the *becoming* knowledge state and *current* knowledge state
2.2.3. Uncertainty	Knowledge incompleteness due to inherent deficiencies with acquired knowledge.
2.2.3.1. Ambiguity	The possibility of having multi-outcomes for processes or systems.
a. Unspecificity	Outcomes or assignments that are not completely defined.
b. Nonspecificity	Outcomes or assignments that are improperly defined.
2.2.3.2. Approximations	A process that involves the use of vague semantics in language, approximate reasoning, and dealing with complexity by emphasizing relevance.
a. Vagueness	Noncrispness of belonging and nonbelonging of elements to a set or a notion of interest.
b. Coarseness	Approximating a crisp set by subsets of an underlying partition of the set's universe that would bound the set of interest.
c. Simplifications	Assumptions needed to make problems and solutions tractable.
2.2.3.3. Likelihood	Defined by its components of randomness, statistical and modeling.
a. Randomness	Nonpredictability of outcomes.
b. Sampling	Samples versus populations.

be attributed to intuitions of experts that could not be negotiated with others in terms of cognitive relevance. Taboo is due to socially reinforced irrelevance. Issues that people must not know, deal with, inquire about, or investigate define the domain of taboo. The undecidability type deals with issues that cannot be designated true or false because they are considered insoluble, or solutions that are not verifiable, or as a result of *ignoratio elenchi*. A third component of blind ignorance is fallacy that can be defined as erroneous beliefs due to misleading notions.

Kurt Gödel (1906–1978) showed that a logical system could not be both consistent and complete and could not prove itself complete without proving itself inconsistent and vise versa. Also, he showed that there are problems that cannot be solved by any set of rules or procedures; instead, for these problems one must always extend the set of axioms. This philosophical view of logic can be used as a basis for classifying the conscious ignorance into *inconsistency* and *incompleteness*.

Inconsistency in knowledge can be attributed to distorted information as a result of inaccuracy, conflict, contradiction, and/or confusion, as shown in Figure 1.9. Inconsistency can result from assignments and substitutions that are wrong, conflicting, or biased, producing confusion, conflict, or inaccuracy, respectively. The confusion and conflict result from in-kind inconsistent assignments and substitutions, whereas inaccuracy results from a level bias or error in these assignments and substitutions.

Incompleteness is defined as incomplete knowledge and can be considered to consist of (1) absence and unknowns as incompleteness in kind and (2) uncertainty. The unknowns or unknown knowledge can be viewed in evolutionary epistemology as the difference between the *becoming* knowledge state and *current* knowledge state. The knowledge absence component can lead to one of these scenarios: (1) no action and working without the knowledge, (2) unintentionally acquiring irrelevant knowledge, leading to blind ignorance, (3) acquiring relevant knowledge that can be with various uncertainties and levels, or (4) deliberately acquiring irrelevant knowledge. The fourth possible scenario is frequently not listed, since it is not realistic.

Uncertainty can be defined as knowledge incompleteness due to inherent deficiencies with acquired knowledge. Uncertainty can be classified into three types based on its sources: ambiguity, approximations, and likelihood. Ambiguity comes from the possibility of having multiple outcomes for processes or systems. Recognition of some of the possible outcomes creates uncertainty. The recognized outcomes might constitute only a partial list of all possible outcomes leading to unspecificity. In this context, unspecificity results from outcomes or assignments that are not completely defined. The incorrect definition of outcomes, i.e., error in defining outcomes, can be called nonspecificity. In this context, nonspecificity results from outcomes or assignments that are improperly defined. Nonspecificity is a form of knowledge absence and can be treated similarly to the absence category under incompleteness. Nonspecificity can be viewed as a state of blind ignorance.

The human mind has the ability to perform approximations through reduction and generalizations, i.e., induction and deduction, respectively, in developing knowledge. The process of approximation can involve the use of vague semantics in language, approximate reasoning, and dealing with complexity by emphasizing relevance. Approximations can be viewed to include vagueness, coarseness, and simplification. Vagueness results from the non-crisp nature of belonging and nonbelonging of elements to a set or a notion of interest, whereas coarseness results from approximating a crisp set by subsets of an underlying partition of the set's universe that would bound the crisp set of interest. Simplifications are assumptions made to make problems and solutions tractable.

Likelihood can be defined in the context of chance, odds, and gambling, and it has primary components of randomness and sampling. Randomness stems from the non-predictability of outcomes. Engineers and scientists commonly use samples to characterize populations, hence the last type.

1.6 Exercise Problems

Problem 1.1. Using Plato's theory of reality, provide three examples of forms or ideas that are not used in this book. Why do they represent higher knowledge levels?

Problem 1.2. What is skepticism? Describe its origin and progression through the times.

Problem 1.3. Write an essay of about 400 words on the book *Tuhafut al-Tuhafut* by Ibn Rushd summarizing its primary arguments, its significance, and its effect on Europe.

Problem 1.4. What is positivism? Describe its origin and progression.

Problem 1.5. What is the theory of meaning?

Problem 1.6. What are the differences between knowledge, information, and opinions?

Problem 1.7. What is ignorance?

Problem 1.8. What are knowledge types, and sources? Provide examples.

Problem 1.9. Provide an engineering example of the Gettier problem.

Problem 1.10. Provide engineering examples of the various ignorance types in the hierarchy provided in Figure 1.9.

Chapter one: Knowledge and ignorance 35

Problem 1.11. Provide an example from the sciences of the Gettier problem.

Problem 1.12. Provide examples from the sciences of the various ignorance types in the hierarchy provided in Figure 1.9.

Problem 1.13. What are the differences between an unknown and an unknowable? Provide examples.

chapter two

Information-based system definition

Contents

- 2.1. Introduction 38
- 2.2. System definition models 42
  - 2.2.1. Perspectives for system definition 42
  - 2.2.2. Requirements and work breakdown structure 43
    - 2.2.2.1. Requirements analysis 43
    - 2.2.2.2. Work breakdown structure 45
  - 2.2.3. Process modeling method 45
    - 2.2.3.1. System engineering process 46
    - 2.2.3.2. Lifecycle of engineering systems 49
    - 2.2.3.3. Technical maturity model 54
  - 2.2.4. Black-box method 54
  - 2.2.5. State-based method 57
  - 2.2.6. Component integration method 58
  - 2.2.7. Decision analysis method 59
    - 2.2.7.1. Decision variables 59
    - 2.2.7.2. Decision outcomes 61
    - 2.2.7.3. Associated probabilities and consequences 61
    - 2.2.7.4. Decision trees 61
    - 2.2.7.5. Influence diagrams 64
- 2.3. Hierarchical definitions of systems 64
  - 2.3.1. Introduction 64
  - 2.3.2. Knowledge and information hierarchy 66
    - 2.3.2.1. Source systems 67
    - 2.3.2.2. Data systems 72
    - 2.3.2.3. Generative systems 75
    - 2.3.2.4. Structure systems 81
    - 2.3.2.5. Metasystems 82

2.4. Models for ignorance and uncertainty types 83
    2.4.1. Mathematical theories for ignorance types 83
    2.4.2. Information uncertainty in engineering systems 85
        2.4.2.1. Abstraction and modeling of engineering systems 85
        2.4.2.2. Ignorance and uncertainty in abstracted aspects of a system 88
        2.4.2.3. Ignorance and uncertainty in nonabstracted aspects of a system 90
        2.4.2.4. Ignorance due to unknown aspects of a system 90
2.5. System complexity 91
2.6. Exercise problems 95

2.1 Introduction

The definition and articulation of problems in engineering and science are critical tasks in the processes of analysis and design and can be systematically performed within a systems framework. "The mere formulation of a problem is often far more essential than its solution," Albert Einstein (1879–1955) said. "What we observe is not nature itself, but nature exposed to our method of questioning," Werner Karl Heisenberg (1901–1976) said. Generally, a manmade or natural system, such as an engineering project, can be modeled to include a segment of its environment that interacts significantly with it to define an underlying system. The boundaries of the system are drawn based on the goals and characteristics of the analysis, the class of performances (including failures) under consideration, and the objectives of the analysis.

    A generalized systems formulation allows scientists and engineers to develop a complete and comprehensive understanding of the nature of a problem, underlying physics, processes, and activities. In a system formulation, an image or a model of an object that emphasizes some important and critical properties is defined. System definition is usually the first step in an overall methodology formulated for achieving a set of objectives. This definition can be based on observations at different system levels that are established based on these objectives. The observations can be about the different elements (or components) of the system, interactions among these elements, and the expected behavior of the system. Each level of knowledge that is obtained about an engineering problem defines a system to represent the project or the problem. As additional levels of knowledge are added to previous ones, higher epistemological levels of system definition and description are attained which, taken together, form a hierarchy of the system descriptions.

    Informally, what is a system? According to the Webster's dictionary, a system is defined as "a regularly interacting or interdependent group of items forming a unified whole." For scientists and engineers, the definition can be stated as "a regularly interacting or interdependent group of items forming a unified whole that has some attributes of interest." Alternatively, a system can be defined as a group of interacting, interrelated, or interdependent elements that together form a complex whole that can be a complex physical structure, process, or procedure of some attributes of interest. All parts of a system are related to the same overall process, procedure, or structure, yet they are most likely different from one another and often perform completely different functions.

In engineering, the discipline of systems engineering establishes the configuration and size of system hardware, software, facilities, and personnel through an interactive process of analysis and design in order to satisfy an operational mission need for the system in a cost-effective manner. A system engineering process identifies mission requirements and translates them into design requirements at succeeding lower levels to insure operational and performance satisfaction. Control of the evolving development process is maintained by a system engineering organization through a continuing series of reviews and audits of technical documentation produced by system engineering and other engineering organizations. The essence of system engineering is structure. Therefore, a system engineer is expected to analyze and define the system as a set of elements or parts so connected to form a whole. System engineers understand the system by bringing structure to it. The particular structure chosen is key to system engineers' understanding by making the needed choices to determine what constitutes its elements and associated technologies, cost, and schedule for the success of the completed system. There are no clearly defined guidelines for the choice of system elements. However, the elements define interfaces among these elements. Understanding, controlling, and optimizing interfaces are a major task of system engineers. More time can be spent on the interfaces than on the elements themselves. System engineers leverage their understanding of the system into developing requirements among the elements. Structured approaches give a mechanistic listing of interaction among the elements. An understanding of the entire system is necessary for a complete listing of all interface requirements. Understanding the "Big Picture" is key to identifying interfaces that affect the elements chosen, which can change the structure of the system. Figure 2.1 shows how system engineers as people identify needs from an environment, define engineering problems, and provide solutions that feed into the environment through a dynamic process.

Example 2.1   Safety of flood-control dams

The primary purposes of most flood-control dams are flood control and grade stabilization. A secondary function is trapping sediment. Flood-control dams are designed and constructed for a sufficient capacity to store runoffs from a ten- to hundred-year storm. A principal spillway is commonly used to pass floodwater from the storage pool, i.e., a dam's reservoir, by means of a pipe through the dam over a period of several days. Any

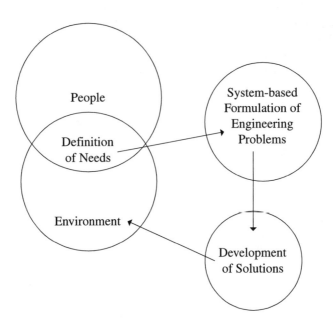

Figure 2.1 Engineers and systems.

excess runoff from a storm passes immediately over an emergency spillway. The emergency spillway is usually a grassed waterway. Some flood control dams in dry and windy areas rarely contain any water but must have large capacities to control flash floods. Figures 2.2 and 2.3 show a flooded dam and a dam failure, respectively. Figure 2.2 shows workers trying to cross a flooded dam, and Figure 2.3 shows a segment of the failed reservoir of the dam.

The U.S. Army Corps of Engineers (USACE) has the responsibility of planning, designing, constructing and maintaining a large number of U.S. flood-control dams. The safety of these dams is of great interest to the USACE. The safety assessment of a dam requires defining a dam system to include (1) the dam facility of structures, foundations, spillways, equipment, warning systems, and personnel, (2) the upstream environment that can produce storms and floods, and (3) the downstream environment that includes the potential sources of flood consequences. Due to the complexity of storm development and yield, the upstream segment of the system is difficult to define and would require a substantial effort level to study. Similarly, the downstream segment is complex in its nature and methods of assessment. The dam facility typically receives the bulk of engineering attention. Systems engineers need to properly define the system with a proper allocation of details in order to achieve an intended study goal.

Figure 2.2 Workers crossing Lacamas Lake Dam in Camas, Washington, during February, 1996 flood (with permission as a courtesy of the Washington State Dam Safety Office).

Figure 2.3 Dam failure on the slope of Seminary Hill, Centralia, Washington, 1991 (with permission as a courtesy of the Washington State Dam Safety Office).

2.2 System definition models

2.2.1 Perspectives for system definition

The term *system* originates from the Greek word *systēma* that means an organized whole. The Webster's dictionary defines it as "a set or arrangement of things so related or connected as to form a unity or organic whole," such as a solar system, school system, or system of highways. Also, it is defined as "a set of facts, principles, rules, etc. classified or arranged in a regular, orderly form so as to show a logical plan linking the various parts." The term *system science* is usually associated with observations, identification, description, experimental investigation, and theoretical modeling and explanations that are associated with natural phenomena in fields, such as biology, chemistry, and physics. *System analysis* includes ongoing analytical processes of evaluating various alternatives in design and model construction by employing mathematical methods, for example, optimization, reliability assessment, statistics, risk analysis, and operations research.

Systems can be grouped in various categories, such as (1) natural systems, such as river systems, and energy systems; (2) manmade systems that can be imbedded in the natural systems, such as hydroelectric power systems and navigation systems; (3) physical systems that are made of real components occupying space, such as automobiles and computers; (4) conceptual systems that could lead to physical systems; (5) static systems that are without any activity, such as bridges subjected to dead loads; (6) dynamic systems, such as transportation systems; (7) closed or open-loop systems, such as a chemical equilibrium process and logistic systems, respectively. Blanchard (1998) provides additional information on these categories.

Systems analysis requires the development of models that need to be representative of system behavior by focusing on selected attributes of interest. Models for various categories, including natural or manmade systems, can be viewed as abstractions of their respective real systems. System scientists or engineers play a major role in defining the level of details of this abstraction, and type and extent of information needed in order to properly and adequately model these attributes and predict system behavior. In general, models can be viewed as assemblages of knowledge and information on most relevant system behavior and attributes. The availability of knowledge and information, lack of them, and uncertainty play major roles in defining these models. This section summarizes various system models that include (1) requirements analysis, (2) work breakdown structure, (3) process modeling method, (4) black-box method, (5) state-based method, (6) component-integration method, and (7) decision-analysis method. It is very common to use a combination of several models to represent a system in order to achieve an analytical goal.

2.2.2 Requirements and work breakdown structure

2.2.2.1 Requirements analysis

The definition of a system requires a goal that can be determined from either need identification or a problem articulation. The goal statement should then be used to define a hierarchy of objectives that can be used to develop a list of performance and functional requirements for the system. Therefore, these requirements form a basis for other system definition methods that are described in subsequent sections.

A system model can be developed based on requirement and functional modeling. For example, dams can be modeled as systems with functional and performance requirements in an environment that has natural and man-made hazards. By limiting the model only to the physical system of a dam, Figure 2.4 was developed for illustrative purposes. The functional requirements of a dam are used to develop a system breakdown. The system breakdown structure is the top down hierarchical division of the dam into its subsystems and components, including people, structure, foundation, flood plain, river and its tributaries, procedures, and equipment. By dividing the dam environment into major subsystems, an organized physical definition for the dam system can be created. This definition allows for a better evaluation of hazards and potential effects of these hazards. By evaluating risk hierarchically (top down) rather than in a fragmented manner, rational, repeatable, and systematic outcomes can be achieved.

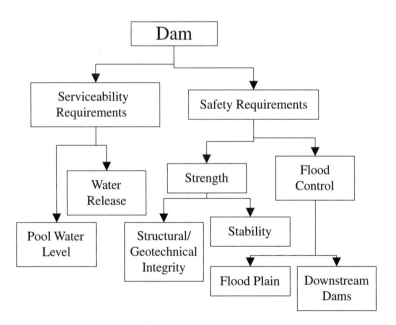

Figure 2.4 Functional requirements for a dam.

Requirements analysis can be defined as the detailed study of the system's performance requirements to ensure that the completed system achieves its intend utility to the customer and meets the goal stated. According to this method, the customer's needs should be determined, evaluated for their completeness, and translated into quantifiable, verifiable, and documented performance requirements. Requirements analysis feeds directly into functional analysis, and allocation, design, and synthesis.

Functional analysis examines the characteristic actions of hardware, software, facilities, or personnel that are needed for the system in order to satisfy performance requirements of the system. Functional analysis might establish additional requirements on all supporting elements of the system by examining their detailed operations and interactions. The overall set of system requirements derived by these analyses lead to both performance and functional requirements. Functional requirements define what the system must do and are characterized by verbs because they imply action on the part of the system. The system gathers, processes, transmits, informs, states, initiates, and ceases. Also, physical requirements might be needed and can be included as a part of the performance requirements. Physical requirements define the system's physical nature, such as mass, volume, power, throughput, memory, and momentum. They may also include details down to type and color of paint, location of the ground segment equipment, and specific environmental protection. For example, aerospace company systems, unlike many commercial products, strongly emphasize the functional requirements, thus prompting the need for a significant evaluation of the system's functional requirements and the allocation of those requirements to the physical architecture.

The functional requirements can be loosely assembled into a hierarchy of functional, sequential, communication, procedural, temporal, or logical manner as follows:

- Functional requirements with subfunctions that contribute directly to performing a single function;
- Sequential breakdowns that show data flow processed sequentially from input to output;
- Communication breakdowns based on information and data needs;
- Procedural breakdowns based on logic flow paths;
- Temporal breakdowns for differing functions at different times; and
- Logical breakdowns based on developing logical flows for functions.

Many programs develop multiple functional hierarchies, using more than one of these criteria to sort and decompose the functions. Each criterion provides a different way of looking at the information, which is useful for solving different types of problems. The most common functional hierarchy is a decomposition based on functional grouping, where the lower tier functions taken in total describe the activity of the upper tier function, providing a more detailed description of their top-level functions.

Chapter two: Information-based system definition

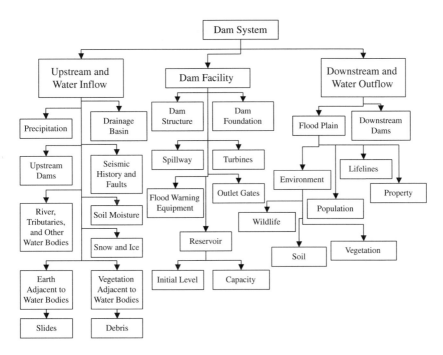

Figure 2.5 Work breakdown structure for a dam.

2.2.2.2 Work breakdown structure

The work breakdown structure, as shown in Figure 2.5 for a dam, is a hierarchy that defines a system's hardware, software, processes, and services. The work breakdown structure is a physically-oriented family tree composed of hardware, software, services, processes, and data that result from engineering efforts during the design and development of a system. The example breakdown for a dam into systems and subsystems of Figure 2.5 focuses on the physical subsystems, components, and the human population at risk. The system was divided into subsystems, such as the dam facility subsystem that includes structural members, foundations, gates, turbines, spillway, and alarms, and reservoir. The work breakdown structure was developed for the goal of performing risk analysis of dams. Each subsystem can be affected by, and can affect other subsystems outside the hierarchy presented. While this breakdown is not complete, it does illustrate the hierarchy of the system and subsystem relations.

2.2.3 Process modeling method

The definition of a system can be viewed as a process that emphasizes an attribute of interest of the system. The steps involved in this process form a spiral of system definitions with hierarchical structure and solutions of problems through decision analysis by learning, abstraction, modeling, and

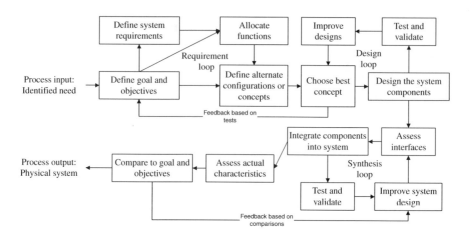

Figure 2.6 System engineering process.

refinement. Example processes include engineering systems as products to meet user demands, engineering systems with lifecycles, and engineering systems defined by a technical maturity process. These three example processes are described in subsequent sections for demonstration purposes.

2.2.3.1 System engineering process

The system engineering process focuses on the interaction between humans and the environment as schematically provided in Figure 2.1. The steps involved in a system engineering process can be viewed to constitute a spiral hierarchy. A system engineering process has the following steps as shown in Figure 2.6:

1. **Recognition of need or opportunity.** The recognition of need or opportunity results from the interaction of humans with various environments, and, therefore, this step can be considered as not being a part of the spiral but its first cause. The step can be viewed as an entrepreneurial activity, rather than an engineering task. The discovery of a need can be articulated in the form of a goal for a proposed system with a hierarchical breakdown into objectives. The delineation of the system's goal should form the basis for and produce the requirements desired by eventual users of the system. For a government, the system's goal should also include the long-term interests of the public.
2. **Identification and qualification of the goal, objectives, and performance and functional requirements.** The goal or mission of the system needs to be stated and must be delineated. This statement should then be used to define a hierarchy of objectives that can be used to develop a list of performance requirements for the systems.

These definitions of the goal, objectives and performance requirements can be used to compare the cost effectiveness of alternative system design concepts. The objectives and performance requirements should include relevant aspects of effectiveness, cost, schedule and risk, and they should be traceable to the goal. To facilitate tradeoff analyses, they should be stated in quantifiable and verifiable terms to some meaningful extents. At each turn of a loop or spiral, the objectives and performance requirements should be documented for traceability to various system components. As the systems engineering process continues, the performance requirements need to be translated into a functional hierarchy for the system allocated to the system's components. The performance and functional requirements should be quantitatively described.

3. **Creation of alternative design concepts.** Establishing a clear understanding of what the system needs to accomplish is a prerequisite to devising a variety of ways that the goal, objectives, and requirements can be met. Sometimes, the alternatives can come about as a consequence of integrating available component design options. Using a bottom-up alternative creation, various concept designs can be developed. It is essential to maintain objectivity to the process without being drawn to a specific option that would limit or obscure the examination of other options. An analyst or designer must stay an outsider in order to maintain objectivity. This detachment would allow the analyst or designer to avoid a premature focus on a single design and would permit discovery of a truly superior design.

4. **Testing and validation.** At this stage, some testing and validation of the concepts might be necessary in order to establish an understanding of the limitations, capabilities, and characteristics of various concepts. The testing and validation can be experimentally, analytically, or numerically performed using laboratory tests, analytical models, or simulation, respectively. The insight gained from this step might be crucial for subsequent steps of this process.

5. **Performance of tradeoff studies and selection of a design.** Tradeoff studies start by assessing how well each design concept meets the system's goal, objectives, and requirements including effectiveness, cost, schedule, and risk, both quantitatively and otherwise. This assessment can utilize the testing and validation results of the previous step. These studies can be performed using system models that analytically relate various concept characteristics to performance and functional requirements. An outcome of these studies can be the determination of bounds on the relative cost effectiveness of the design concepts. Selection among the alternative design concepts must take into account subjective factors that are not quantifiable and were not incorporated in the studies. When possible, mathematical expressions, called objective functions, should be developed and used to express the values of combinations of possible outcomes as

a single measure of cost effectiveness. The outcome of this step is to identify the best concept that should be advanced to next steps.

6. **Development of a detailed design.** One of the first issues to be addressed is how the system should be subdivided into subsystems and components in order to represent accurately an engineering product of interest. The partitioning process stops when the subsystems or components are simple enough to be managed holistically. Also, the system might reside within a program that has well-established activities or groups. The program's activities might drive the definitions of the system hierarchy of subsystems and components. These program activities should be minimized in number and complexity as they define various interfaces, and they could have a strong influence on the overall system cost and schedules. Partitioning is more of an art than a science, however, experiences from other related systems and judgment should be utilized. Interfaces can be simplified by grouping similar functions, designs and technologies. The designs for the components and subsystems should be tested, verified and validated. The components and subsystems should map conveniently onto an organizational structure if applicable. Some of the functions that are needed throughout the system (such as electrical power availability) or throughout the organization (such as purchasing) can be centralized. Standardization of such things as parts lists or reporting formats is often desirable. The accounting system should follow, not lead, the system architecture. Partitioning should be done essentially all at once, broadly covering the entire system. Similar to system design choices, alternative partitioning plans should be considered and compared before selecting the optimal plan and its implementation.

7. **Implementing the selected design decisions.** The design spiral or loop of successive refinement should proceed until reaching diminishing returns. The next step is to reverse the partitioning process by unwinding the process. This unwinding phase is called *system integration*. Conceptual system integration takes place in all steps of the process, i.e., when a concept has been selected, the approach is verified by unwinding the process to test whether the concept at each physical level meets the expectations and requirements. The physical integration phase is accomplished during fabrication or manufacturing of the system. The subsystem integration should be verified and validated to ensure that the subsystems conform to design requirements individually and at the interfaces, such as mechanical connections, power consumption, and data flow. System verification and validation consist of ensuring that subsystems interfaced achieve their intended results collectively as one system.

8. **Performance of missions.** In this step, the physical system is called upon to meet the need for which it was designed and built. During this step, the system effectiveness at the operational site needs to be validated. Also, this step includes maintenance and logistics

Chapter two: Information-based system definition   49

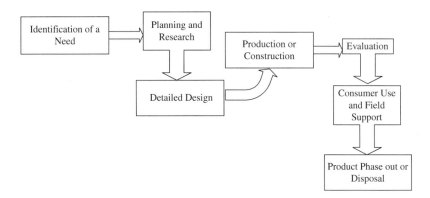

Figure 2.7 Lifecycle of engineering systems.

documentation, definition of sustaining engineering activities, compilation of development and operations *lessons learned* documents, and, with the help of the specialty engineering disciplines, identification of product improvement opportunities for quantifiable system objectives. Sometimes only bounds, rather than final values, are possible in this step. The spread between any upper- and lower-bound estimates of system attributes or performances can be reduced as a result of increasing the level of validation and testing, and continually improving and enhancing the design.

2.2.3.2 *Lifecycle of engineering systems*

Engineering products can be treated as systems that have a lifecycle. A generic lifecycle of a system begins with the initial identification of a need and extends through planning, research, design, production or construction, evaluation, consumer use, field support, and ultimately product phaseout or disposal, as shown in Figure 2.7.

A system lifecycle is sometimes known as the consumer-to-consumer cycle that has major activities applicable to each phase of the lifecycle, as illustrated in Table 2.1. The steps illustrated show a logical flow and associated functions for each step or effort. Although the generic steps are the same, various systems might require different specific details in terms of "what has to be done." A large system requiring new development, such as a satellite or major ground system, may evolve through all the steps, whereas a relatively small item, such as an element of a space segment or the maintenance phase of a software contract, may not. In considering the lifecycle of a system, each of the steps identified should be addressed even though all steps may not be applicable.

The lifecycle of a product is a general concept that needs to be tailored for each user or customer. The lifecycle of systems according to the Department of Defense (DoD) and the National Aeronautics and Space Administration

Table 2.1 The Consumer-to-Consumer Cycle

Phase	Activities
The System Lifecycle	
Consumer	
Identification of need	"Wants or desires" for systems are because of obvious deficiencies/problems or are made evident through basic research results.
Producer	
System planning function	Marketing analysis; feasibility study; advanced system planning through system selection, specifications and plans, acquisition plan research/design/production, evaluation plan, system use and logistic support plan; planning review; proposal.
System research function	Basic research; applied research based on needs; research methods; results of research; evolution from basic research to system design and development.
System design function	Design requirements; conceptual design; preliminary system design; detailed design; design support; engineering model/prototype development; transition from design to production.
Production and/or construction function	Production and/or construction requirements; industrial engineering and operations analysis such as plant engineering, manufacturing engineering, methods engineering, and production control; quality control; production operations.
Consumer	
System evaluation function	Evaluation requirements; categories of test and evaluation; test preparation phase including planning and resource requirements; formal test and evaluation; data collection, analysis, reporting, and corrective action; retesting.
System use and logistic support function	System distribution and operational use; elements of logistics and lifecycle maintenance support; system evaluation; modifications, product phase-out; material disposal, reclamation, and recycling.

(NASA) are tied to the government procurement process, as discussed in the NASA Example 2.1, but the general applicability of the concept of a system lifecycle is independent of the user and the procurement process.

Example 2.1   Lifecycle of NASA engineering systems

The National Aeronautics and Space Administration (NASA) uses the concept of lifecycle for a program, called a program lifecycle which consists of distinct phases separated by control gates. NASA uses its lifecycle model not only to describe how a program evolves over time, but also to aid management in program control. The boundaries between phases are defined so that they provide *proceed* or *do not proceed* decisions. Decisions to proceed may be qualified by liens that must be removed within a reasonable time. A program that fails to pass a control gate and has enough resources may be allowed to readdress the deficiencies or it may be terminated.

The government operates with a fiscal budget and annual funding leading to implicit funding control gates at the beginning of each fiscal year. While these gates place planning requirements on the project and can make significant replanning necessary, they are not part of an orderly system engineering process; rather, they constitute one of the sources of uncertainty that affect project risks and should be included in project risk considerations.

The NASA model can generally be defined to include the following phases:

Prephase A — Advanced Studies

The objective of this phase is to produce a broad spectrum of ideas and alternatives for missions from which new projects or programs can be selected. Major activities and their products in this prephase A are to (1) identify missions consistent with the NASA charter, (2) identify and involve users, and (3) perform preliminary evaluations of possible missions. Typically, this phase consists of loosely structured examinations of new ideas, usually without central control and mostly oriented toward small studies. Also, program or project proposals are prepared that include mission justification and objectives, possible operations concepts, possible system architectures, and cost, schedule, and risk estimates. The phase also produces master plans for existing program areas. The control gate for this phase is informal proposal reviews. Descriptions of projects suggested generally include initial system design and operational concepts, preliminary project organization, schedule, testing and review structure, and documentation requirements.

This phase is of an ongoing nature since technological progress makes possible missions that were previously impossible. Manned trips to the moon and the taking of high-resolution pictures of planets and other objects in the universe illustrate past responses to this kind of opportunity. New opportunities will continue to become available as our technological capabilities grow.

Phase A — Conceptual design studies

The objective of this phase is to determine the feasibility and desirability of a suggested new major system in preparation for the seeking of funding. This phase includes major activities such as (1) preparation of mission needs statements, (2) development of preliminary system requirements, (3) identification of alternative operations and logistics concepts, (4) identification of project constraints and system boundaries, (5) consideration of alternative design concepts, and (6) demonstrating that credible, feasible designs exist. System validation plans are initiated in this phase. Also, systems engineering tools and models are acquired, environmental impact studies are initiated, and program implementation plans are prepared. The control gates are conceptual design review and prephase B nonadvocate review. This phase is frequently described as a structured version of the previous phase.

Phase B — Concept definition

The objective of this phase is to define the project in enough detail to establish an initial baseline. This phase includes major activities such as

- reaffirmation of the mission needs statement,
- preparation of a program initiation agreement,
- preparation of a system engineering management plan,
- preparation of a risk management plan,
- initiation of configuration management,
- development of a system-level cost-effectiveness model,
- restatement of the mission needs as system requirements,
- establishment of the initial requirements traceability matrix,
- selection of a baseline system architecture at some level of resolution and concept of operation,
- identification of science payload,
- definition of internal and external interface requirements,
- definition of the work breakdown structure,
- definition of verification approach and policies,
- preparation of preliminary manufacturing plans,
- identification of government resource requirements,
- identification of ground test and facility requirements,
- development of statement of work,
- revision and publication of project implementation plans,
- initiation of advanced technology development programs.

The control gates include project definition and cost review, program and project requirements review, and safety review.

Tradeoff studies in this phase should precede rather than follow system design decisions. A feasible system design can be defined as a design that can be implemented as designed, and can then accomplish the system's goal within the constraints imposed by the fiscal and operating environment. To be credible, a design must not depend on the occurrence of unforeseen breakthroughs in the state of the art. While a credible design may assume likely improvements in the state of the art, it is nonetheless riskier than one that does not.

Phase C — Design and development

This phase has the objective to design a system and its associated subsystems, including its operations systems, so that it will be able to meet its requirements. This phase has primary tasks and activities that include

- adding subsystem design specifications to the system architecture,
- publishing subsystem requirements documents,
- preparation of subsystem verification plans,
- preparation of interface documents,
- repetition of the process of successive refinement to get "design-to" and "build-to" specifications and drawings, verification plans, and interface documents at all levels,
- augmentation of documents to reflect the growing maturity of the system,
- monitoring the project progress against project plans,
- development of the system integration plan and the system operations plans,
- documentation of tradeoff studies performed,
- development of the end-to-end information system design and the system deployment approach,
- identification of opportunities for preplanned product improvement, and
- confirmation of science payload selection.

Control gates include system-level preliminary design review, subsystem (and lower level) preliminary design reviews, subsystem (and lower level) critical design reviews, and system-level critical design review.

The purpose of this phase is to unfold system requirements into system and subsystem designs. Several popular approaches can be used in the unfolding process, such as the code-and-fix, the waterfall, requirements-driven design, and/or evolutionary development.

Phase D — Fabrication, integration, test and certification

The purpose of this phase is to build the system designed in the previous phase. Activities include fabrication system of hardware and coding of software, integration, verification and validation, and certified acceptance of the system.

Phase E — Pre-operations

The purpose of this phase is to prepare the certified system for operations by performing main activities that include the initial training of operating personnel and finalization of the integrated logistics support plan. For flight projects, the focus of activities then shifts to prelaunch integration and launch. On the other hand, for large flight projects, extended periods of orbit insertion, assembly, and shakedown operations are needed. In some projects, these activities can be treated as minor items permitting combining this phase with either its predecessor or its successor. The control gates are launch readiness reviews, operational readiness reviews, and safety reviews.

Phase F — Operations and disposal

The objective of this phase is to actually meet the initially identified need, and then to dispose of the system in a responsible manner. This phase includes major activities such as (1) training replacement operators, (2) conducting the mission, (3) maintaining the operating system, and (4) disposing of the system. The control gates are operational acceptance review, regular system operations reviews, and system upgrade reviews.

Phase F encompasses the problem of dealing with the system when it has completed its mission. The end of life depends on many factors. For example, the disposal of a flight system with short-mission duration, such as a space-lab payload, may require only selection and retrieval of the hardware and its return to its owner, whereas a large flight project of long duration disposal may proceed according to long-established plans or may begin as a result of unplanned events, such as accidents. In addition to uncertainty as to when this part of the phase begins, the activities associated with safely deactivating and disposing of a system may be long and complex. As a result, the costs and risks associated with different designs should be considered during the planning process.

2.2.3.3 Technical maturity model

The technical maturity model is another view of the lifecycle of a project. According to this model, the lifecycle considers a program as an interaction between society and engineering. The model concentrates on the engineering aspects of the program and not on the technology development through research. The program must come to fruition by meeting both the needs of the customer and also meeting the technical requirements. Therefore, by keeping distinctions among technical requirements, needs and technology development, the motivations, wants, and desires of the customer are differentiated from the technology issues during the course of the project.

2.2.4 Black-box method

Historically, engineers have built analytical models to represent natural and manmade systems using empirical tools of observing system attributes of interest (called system output variables) and trying to relate them to some other controllable or uncontrollable input variables. For example, a structural engineer might observe the deflection of a bridge as an output of an input such as a load at middle of its span. By varying the intensity of the load, the deflection changes. Empirical test methods would vary the load incrementally and the corresponding deflections are measured, thereby producing a relationship such as

$$y = f(x) \tag{2.1}$$

where x = input variable, y = output variable, and f = a function that relates input to output. In general, a system might have several input variables that can be represented as a vector X, and several output variables that can be represented by a vector Y. A schematic representation of this model is shown in Figure 2.8. According to this model, the system is viewed as a whole entity without any knowledge on how the input variables are processed within the system to produce the output variables. This black box view of the system has the advantage of shielding an analyst from the physics governing the system, and providing the analyst with the opportunity to focus on relating the output to the input within some range of interest for the underlying variables. The primary assumptions according to this model are (1) the existence of causal relationships between input and output variables as defined by the function f, and (2) the effects of time, i.e., time-lag or time-prolongation within the system, are accounted for by methods of measurement of input and output variables.

For complex engineering systems or natural systems, the numbers of input and output variables might be large with varying levels of importance. In such cases, a system engineer would be faced with the challenge of identifying the most significant variables, and how they should be measured. Establishing a short list of variables might be a most difficult task especially for novel systems. Some knowledge of the physics of the system might help in this task of system identification. Then, the analyst needs to decide on the nature of time-relation between input and output by addressing questions such as

- Is the output instantaneous as a result of the input?
- If the output lags behind the input, what is the lag time? Are the lag times for the input and output related, i.e., exhibiting nonlinear behavior?
- Does the function f depend on time, number of input applications, or magnitude of input?
- Does the input produce an output, and linger within the system affecting future outputs?

These questions are important for the purpose of defining the model, its applicability range, and validity.

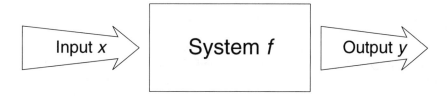

Figure 2.8 Black-box system model.

Example 2.2 Probable maximum flood

The U.S. Army Corps of Engineers (USACE) classes dams according to both size and hazard, where hazard is defined in terms of potential loss of life and economic loss (Committee on Safety Criteria for Dams, 1985). Small dams are 25 to 40 ft high, intermediate dams are 40 to 100 ft high, and large dams are over 100 ft high. Low hazard dams are those for which failure of the dam would result in no loss of life and only minimal economic loss. A significant hazard is one that would cause a few losses of life and appreciable economic loss, and a high hazard would result in the loss of more than a few lives and excessive economic loss.

The USACE uses three methods of determining extreme floods, depending on the return period and intended use (USACE, 1965). Frequency analyses are used when the project demands a storm event with a relatively common return period and are based on gage records. This type of analysis is used for low hazard dams, small to intermediate dams in size, or small dams with significant hazard classifications. A standard project flood (SPF) is used when some risk can be tolerated but where an unusually high degree of protection is justified because of risk to life and property (Ponce, 1989). The SPF includes severe combinations of meteorological and hydrological conditions but does not include extremely rare combinations. The SPF is typically used for dams that are classed as a significant hazard and intermediate to large in size. For projects requiring substantial reduction in risk, such as dams classed as a high hazard, the probable maximum flood (PMF) is used. The PMF is caused by the most severe and extreme combination of meteorological and hydrological events that could possibly occur in an area. Flood prediction can be based on black-box models as shown in Figure 2.9. For river systems, time can play a major role in the form of time lag, time prolongation, and system nonlinearity.

Frequency analyses of gaged data conducted by the USACE are based on recommendations in Bulletin 17B (U.S. Interagency Advisory Committee on Water Data, 1982). The SPF is developed from a standard project storm. The PMF is based on an index rainfall and a depth-area-duration relationship. A hydrograph is then developed based on this rainfall minus hydrologic extractions. For basins less than 1000 mi^2 (2590 km^2), the storms are usually based on localized thunderstorms; for basins greater than 1000 mi^2 (2590 km^2), the storms are usually a combination of events. Due to these differences, the PMF for the smaller basins is based on a 6-hr or 12-hr time increment. For large basins, this procedure is considerably more complex. The SPF is developed very similarly to the PMF except that the index flood is decreased by about 50%.

The use of the PMF has often been questioned since rainfalls and floods of that magnitude have not been experienced in a lifetime. However, studies conducted by the USACE have shown that dozens of storms across the U.S. have exceeded 1/2 of the probable maximum precipitation (PMP) for that area (USACE, 1982; Committee on the Safety of Existing Dams, 1983). Based

Chapter two: Information-based system definition

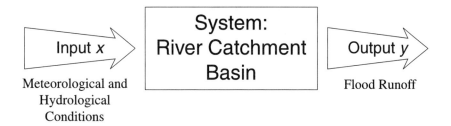

Figure 2.9 Black-box system model for flood prediction.

on these data, the USACE assumes that the PMP is a reasonable basis to estimate the maximum likely hydrological event, although it continued to be debated by its engineers.

2.2.5 State-based method

A convenient modeling method of systems can be based on identifying state variables that would be monitored either continuously or at discrete times. The values of these state variables over time provide a description of the model needed for a system. The state variables should be selected such that each one provides unique information. Redundant state variables are not desirable. The challenge faced by system engineers is to identify the minimum number of state variables that would accurately represent the behavior of the system over time. Also, there is a need to develop models that describe the transitions of state variables from some values to another set of values. It is common that these transitions are not predictable due to uncertainty and can only be characterized probabilistically. The sate transition probabilities are of interest and can be empirically assessed and modeled using, for example, Markov chains for modeling the reliability of repairable systems (Kumamoto and Henley, 1996) as described in Example 2.3.

Example 2.3 Markov modeling of repairable systems

Repairable systems can be assumed for the purpose of demonstration to exist in either a normal (operating) state or a failed state, as shown in Figure 2.10. A system in a normal state makes transitions to either normal states that are governed by its reliability level (i.e., it continues to be normal) or to the failed states through failure. Once it is in a failed state, the system makes transitions to either failed states that are governed by its repair-ease level (i.e., it continues to be failed) or to the normal states through repair. Therefore, four transition probabilities are needed for the following cases:

- Normal-to-normal state transition
- Normal-to-failed state transition

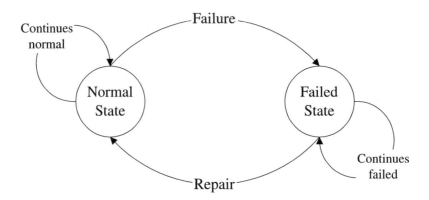

Figure 2.10 A Markov transition diagram for repairable systems.

- Failed-to-failed state transition
- Failed-to-normal state transition

These probabilities can be determined by testing the system and/or by analytical modeling of the physics of failure and repair logistics as provided by Kumamoto and Henley (1996).

2.2.6 Component integration method

Systems can be viewed as assemblages of components. For example, in structural engineering a roof truss can be viewed as a multicomponent system. The truss in Figure 2.11 has 13 members. The principles of statics can be used to determine member forces and reactions for a given set of joint loads. By knowing the internal forces and material properties, other system attributes such as deformations can be evaluated. In this case, the physical connectivity of the real components can be defined as the connectivity of the components in the structural analysis model. However, if we were interested

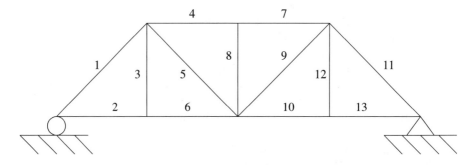

Figure 2.11 A truss structural system.

Chapter two: Information-based system definition

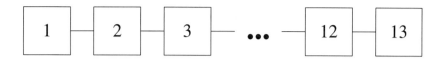

13 components

Figure 2.12 A system in series for the truss as a reliability block diagram.

in the reliability and/or redundancy of the truss, a more appropriate model would be as shown in Figure 2.12, called a reliability block diagram. The representation of the truss in Figure 2.12 emphasizes the attributes of reliability or redundancy. According to this model, the failure of one component would result in the failure of the truss system. Ayyub and McCuen (1997), Ang and Tang (1990), and Kumamoto and Henley (1996) provide details on reliability modeling of systems.

2.2.7 *Decision analysis method*

The elements of a decision model need to be constructed in a systematic manner with a decision-making goal or objectives for a decision-making process. One graphical tool for performing an organized decision analysis is a decision tree, constructed by showing the alternatives for decision-making and associated uncertainties. The result of choosing one of the alternative paths in the decision tree is the consequences of the decision (Ayyub and McCuen, 1997).

The construction of a decision model requires the definition of the following elements: objectives of decision analysis, decision variables, decision outcomes, and associated probabilities and consequences. The objective of the decision analysis results in identifying the scope of the decisions to be considered. The boundaries for the problem can be determined from first understanding the objectives of the decision-making process and using them to define the system.

2.2.7.1 *Decision variables*

The decision variables are the feasible options or alternatives available to the decision maker at any stage of the decision-making process. The decision variables for the decision model need to be defined. Ranges of values that can be taken by the decision variables should be defined. Decision variables in inspecting mechanical or structural components in an industrial facility can include what and when to inspect components or equipment, which inspection methods to use, assessing the significance of detected damage, and repair/replace decisions. Therefore, assigning a value to a decision variable means making a decision at a specific point within the process. These points within the decision-making process are called *decision nodes*, which are identified in the model by a square, as shown in Figure 2.13.

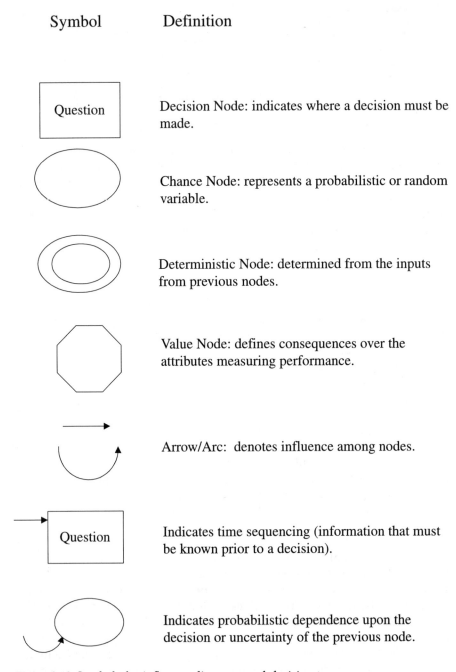

Figure 2.13 Symbols for influence diagrams and decision trees.

2.2.7.2 Decision outcomes

The decision outcomes for the decision model need also to be defined. The decision outcomes are the events that can happen as a result of a decision. They are random in nature, and their occurrence cannot be fully controlled by the decision maker. Decision outcomes can include the outcomes of an inspection (detection or nondetection of a damage) and the outcomes of a repair (satisfactory or nonsatisfactory repair). Therefore, the decision outcomes with the associated occurrence probabilities need to be defined. The decision outcomes can occur after making a decision at points within the decision-making process called *chance nodes*, which are identified in the model using a circle, as shown in Figure 2.13.

2.2.7.3 Associated probabilities and consequences

The decision outcomes take values that can have associated probabilities and consequences. The probabilities are needed due to the random (chance) nature of these outcomes. The consequences can include, for example, the cost of failure due to damage that was not detected by an inspection method.

2.2.7.4 Decision trees

Decision trees are commonly used to examine the available information for the purpose of decision making. The decision tree includes the decision and chance nodes. The decision nodes, that are represented by squares in a decision tree, are followed by possible actions (or alternatives, A_i) that can be selected by a decision maker. The chance nodes, that are represented by circles in a decision tree, are followed by outcomes that can occur without the complete control of the decision maker. The outcomes have both probabilities (P) and consequences (C). Here the consequence can be cost. Each tree segment followed from the beginning (left end) of the tree to the end (right end) of the tree is called a branch. Each branch represents a possible scenario of decisions and possible outcomes. The total expected consequence (cost) for each branch could be computed. Then the most suitable decisions can be selected to obtain the minimum cost. In general, utility values can be used and maximized instead of cost values. Also, decisions can be based on risk profiles by considering both the total expected utility value and the standard deviation of the utility value for each alternative. The standard deviation can be critical for decision-making as it provides a measure of uncertainty in utility values of alternatives (Kumamoto and Henley, 1996). Influence diagrams can be constructed to model dependencies among decision variables, outcomes, and system states using the same symbols of Figure 2.13. In the case of influence diagrams, arrows are used to represent dependencies between linked items, as described in Section 2.2.7.5.

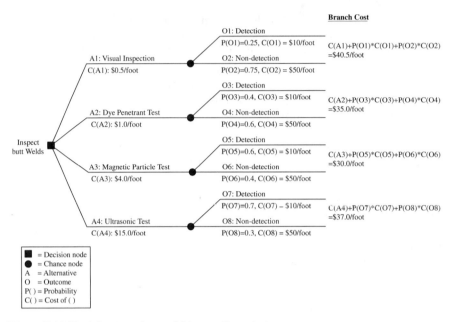

Figure 2.14 Decision tree for weld inspection strategy.

Example 2.4 Decision analysis for selecting an inspection strategy

The objective herein is to develop an inspection strategy for the testing of welds using a decision tree. This example is for illustration purposes and is based on hypothetical probabilities, costs, and consequences.

The first step of the decision analysis for an inspection strategy selection is to identify a system with a safety concern, based on methods, such as risk assessment techniques. After performing the risk assessment, managers must examine various inspection alternatives and select an optimal solution. For example, the welds of a ship's hull plating could be selected as a ship's hull subsystem requiring risk-based inspection. If the welds would fail due to poor weld quality, the adverse consequence could be very significant in terms of economic losses, environmental damages, and potentially human losses and even vessel loss. An adequate inspection program is needed to mitigate this risk and keep it at an acceptable level. Previous experiences and knowledge of the system can be used to identify candidate inspection strategies. For the purpose of illustration, only four candidate inspection strategies are considered, as shown in Figure 2.14: visual inspection, dye penetrant inspection, magnetic particle inspection, and ultrasonic testing.

The outcome of an inspection strategy is either detection or nondetection of a defect, which is identified by an occurrence probability P. These outcomes originate from a chance node. The costs or consequences of these outcomes are identified with the symbol C. The probability and cost estimates were assumed for each inspection strategy based on its portion of the decision tree.

Chapter two: Information-based system definition

The total expected cost for each branch was computed by summing the products of the pairs of cost and probability along the branch. Then total expected cost for the inspection strategy was obtained by adding up the total expected costs of the branches on its portion of the decision tree. Assuming that the decision objective is to minimize the total expected cost, then the "magnetic particle test" alternative should be selected as the optimal strategy. Although this is not the least expensive testing method, its total branch cost is the least. This analysis does not consider the standard deviation of the total cost in making optimal selection. Risk profiles of the candidate inspection strategies can be constructed as the cumulative distribution functions of the total costs for these strategies. Risk dominance can then be identified and an optimal selection can be made.

Example 2.5 Decision analysis for selection of a personal flotation device type

Decision analysis may also be applied to engineered consumer products such as personal flotation devices (PFD's). One application is the assessment of alternative PFD designs based on their performances.

For this example, the objective of the decision analysis is to select the best PFD type based on a combination of the probability of PFD effectiveness and reliability. Probability values were not included as this example is intended to demonstrate only the possible framework for the decision tree as shown in Figure 2.15. The decision criteria could vary based on the

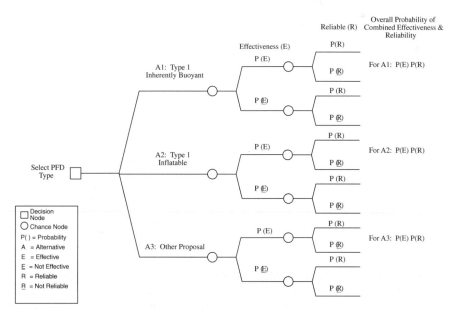

Figure 2.15 Selecting a personal flotation device (PFD) based on effectiveness and reliability.

performance considerations or concerns of the decision maker. For this example, the alternative with the largest value of combined effectiveness and reliability would be the best alternative.

2.2.7.5 Influence diagrams

An influence diagram is a graphical tool that shows the dependence relationships among the decision elements of a system. It is similar to a decision tree; however, influence diagrams provide compact representations of large decision problems by focusing on dependencies among various decision nodes, chance nodes, and outcomes. These compact representations help facilitate the definition and scope of a decision prior to lengthy analysis. They are particularly useful for problems with a single decision variable and a significant number of uncertainties (ASME, 1993). Symbols used for creating influence diagrams are shown in Figure 2.13. Generally, the process begins with identifying the decision criteria and then further defining what influences the criteria. An example of an influence diagram for selecting weld inspection decision criteria is shown in Figure 2.16. An influence diagram showing the relationship of the factors influencing the selection of a personal flotation device (PFD) type is shown in Figure 2.17.

2.3 Hierarchical definitions of systems

2.3.1 Introduction

Using one of the perspectives and models of Section 2.2 to define a system, information then needs to be gathered to develop an *information-based system* definition. The information can be structured in a hierarchical manner to facilitate its construction, completeness, and accuracy of representation, although the resulting hierarchy might not achieve all these requirements. The resulting information structure can be used to construct knowledge levels on the system for the purpose of analyzing and interpreting system behavior. Also, the resulting hierarchy can be used to develop a *generalized system* definition that can generically be used in representing other systems and problems.

A generalized system formulation allows researchers and engineers to develop a complete and comprehensive understanding of manmade products, natural systems, processes, and services. In a system formulation, an *image* or a *model* of an object that emphasizes certain important and critical properties is defined. Systems are usually identified based on the *level of knowledge* and/or information that they contain. Based on their knowledge levels, systems can be classified into five consecutive hierarchical levels. The higher levels include all the information and knowledge introduced in the lower ones in addition to more specific information. System definition is usually the first step in an overall methodology formulated for achieving a set of objectives that defines a goal. For example, in construction management, real-time control of construction or production activities can be one

Chapter two: Information-based system definition 65

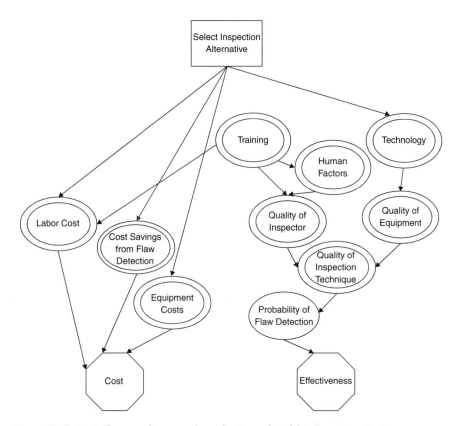

Figure 2.16 An influence diagram for selection of weld inspection strategy.

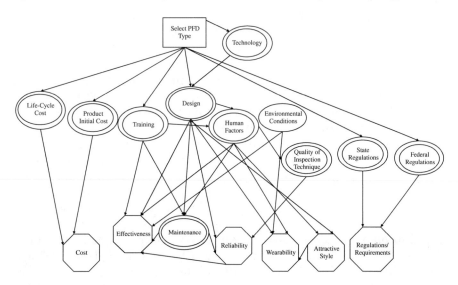

Figure 2.17 An influence diagram for selecting a Personal Flotation Device (PFD) design.

of these objectives. However, in order to develop a control system for a construction activity, this activity has to be suitably defined depending on its nature and methods of control. *Hierarchical control systems* were determined to be suitable for construction activities (Abraham et al., 1989). Thus, the hierarchical nature of a construction activity needs to be emphasized. The generalized system definition as discussed in this section can be used for this purpose. The hierarchical system classification enables the decomposition of the overall construction activity into subsystems that represent the different processes involved in each activity. Then, each process could be decomposed into tasks that are involved in performing the process. Therefore, a breakdown needed for a hierarchical control system is obtained. In this section, basic concepts of system identification and definitions are introduced, together with some additional concepts that could be used in modeling and solving problems in engineering and sciences. Construction activities are modeled and discussed using the methods presented in this section in a systems framework for the purpose of demonstration. The knowledge system is upgraded throughout the course of the coverage in this section from one system level to the next level in order to illustrate the use of the developed concepts for controlling construction activities.

2.3.2 *Knowledge and information hierarchy*

The definition of a system is commonly considered the first step in an overall methodology formulated for achieving a set of objectives (Chestnut, 1965; Hall, 1962 and 1989; Klir, 1969 and 1985; Wilson, 1984). The definition of a system can be interpreted in many ways as discussed in Section 2.2; however, herein a universal definition is used as an arrangement of elements with some important properties and interrelations among them. In order to introduce a comprehensive definition of a system, a more specific description is required based on several main knowledge levels (Klir, 1969 and 1985). Further classifications of systems are possible within each level using methodological distinctions based on, for example, their nature as natural, designed, human activity, and social and cultural (Wilson, 1984). Chestnut (1965) and Hall (1962 and 1989) provided hierarchical formulations of systems based on available information and its degree of detail. Klir (1969 and 1985) introduced a set approach for the system definition problem that was criticized by Hall (1989) because of its inability to express the properties of the overall system, knowing the qualities of its elements. However, for construction activities, the set approach is suitable for representing the variables of the problem. The ability to infer information about the overall system, knowing the behavior of its components, can be dealt with using special techniques as discussed by Klir (1985). Once a system is defined, the next step is to define its environment (Chestnut, 1965; Hall, 1962 and 1989; Klir, 1969 and 1985; Wilson, 1984). The environment is defined as everything within a certain universe that is not included in the system. Hall (1989) introduced an interesting notion within systems thinking that allows the change in boundaries between a defined system and its environment. For the purposes of this section, the formation and structuring of systems are based on the concepts and approaches introduced by Klir (1969 and 1985). The set theory approach serves the objectives of this book well and also the examples presented in this chapter on defining a control system for construction activities. In addition, the approach is formulated in a nonspecific general format and is well suited for computer implementation. In the following sections, knowledge and an example control system are gradually built up in successive levels. Each knowledge level is discussed in detail together with any classifications within each level and an illustrative example.

2.3.2.1 Source systems

At the first level of knowledge, which is usually referred to as level (0), the system is known as a *source system*. Source systems comprise three different components, namely *object systems, specific image systems,* and *general image systems* as shown in Figure 2.18. The object system constitutes a model of the original object. It is composed of an object, attributes, and a backdrop. The object represents the specific problem under consideration. The attributes are the important and critical properties or variables selected for measurement or observation as a model of the original object. The backdrop is the domain, space, within which the attributes are observed. The specific image system is developed based on the object. This image is built through observation channels that measure the attribute variation within the backdrop. The attributes, when measured by these channels, correspond to the variables in the specific image system. The attributes are measured within a support set that corresponds to the backdrop. The support can be time, space, or population. Combinations of two or more of these supports are also possible.

Before upgrading the system to a higher knowledge level, the specific image system can be abstracted into a general format. A mapping function is utilized for this purpose among the different states of the variables to a general state set that is used for all the variables. There are some methodological distinctions that could be defined in this level. Ordering is one of these distinctions that is realized within state or support sets. Any set can be either ordered or not ordered, and those that are ordered may be partially ordered or linearly ordered. An ordered set has elements that can take, for example, real values, or values on an interval or ratio scale. A partially ordered set has elements that take values on an ordinal scale. A nonordered set has components that take values on a nominal scale. Distance is another form of distinction, where the distance is a measure between pairs of elements of an underlying set. It is obvious that if the set is not ordered, the concept of distance is not valid. Continuity is another form of distinction, where variables or support could be discrete or continuous. The classification of the variables as input or output variables forms another distinction. Those systems that have classified input/output variables are referred to as

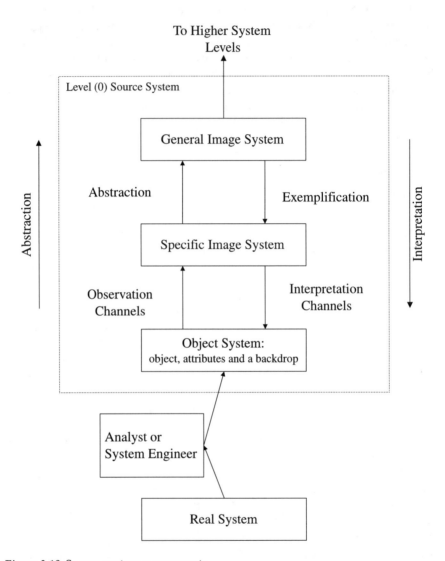

Figure 2.18 Source system components.

directed systems; otherwise they are referred to as neutral systems. The last distinctions that could be realized in this level are related to the observation channels, which could be classified as crisp or fuzzy. Figure 2.19 summarizes methodological distinctions realized in the first level of knowledge. It should be noted that any variation or change in the methodological distinctions of a certain system does not affect its knowledge level.

Chapter two: Information-based system definition

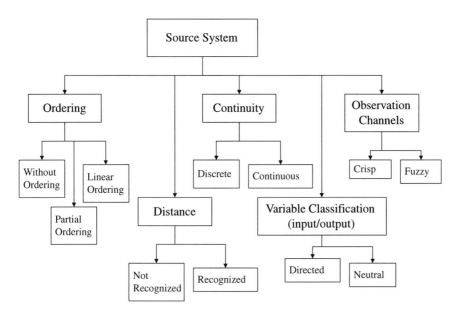

Figure 2.19 Methodological distinctions of source systems.

Example 2.6 *A source system for construction activities*

For the purpose of illustration, the construction activities of concrete placement are considered and their knowledge level upgraded throughout the course of this section. The first step in defining the system for these construction activities is to identify a goal such as construction control by safely placing high quality concrete efficiently and precisely. This goal can be defined through some properties or attributes of interest that can include, for example, safety, quality, productivity, and precision. Considering only two attributes of construction activities, i.e., safety and quality, the variables or factors that affect those attributes should be identified. Only two variables are assumed to affect the safety attribute. These variables could be quantitatively or qualitatively defined depending on their nature. For qualitative variables, fuzzy set theory is used in defining the potential states together with a suitable observation channel that yields a quantitative equivalent for each state (Klir, 1985; Klir and Folger, 1988; Zimmerman, 1985). As an example for this type of variable, labor experience (v_1) is considered. This variable is assumed to have four potential states, namely, fair, good, moderate, and excellent. These linguistic measures can be defined using fuzzy sets. Using a scale of 0 to 10 for the level of experience, these measures can be defined as shown in Figure 2.20. The vertical axis in the figure represents the degree

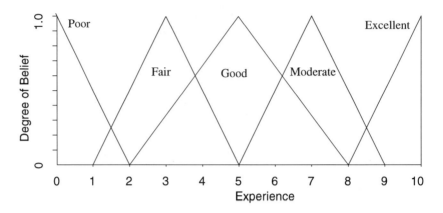

Figure 2.20 Fuzzy definitions of experience.

of belief that the corresponding experience value belongs to the fuzzy sets of Fair, Good, Moderate or Excellent experience, where experience is on a scale of 0 to 10, 0 = absolutely no experience, and 10 = the absolute highest experience. (Fuzzy sets are formally introduced in Chapter 4.) A maximum operator can then be defined in order to get a quantitative equivalent for each state. A one-to-one mapping function is used in order to define the corresponding general states of the variable (v_1). The second variable (v_2) is the method of construction. This variable has, for example, three potential states, namely traditional method, slip-form method, and precast element method. This is a crisp variable, and its observation channel is represented by an engineer who decides which method should be used. A similar one-to-one mapping function is used to relate the different construction methods to the corresponding general states of the variable (v_2).

The next step in the definition of this system, is the identification of the different supports. In this example, the supports include time, space, and population. The time support is needed in measuring the progress of the different variables during the construction period. Assuming a construction period of two months with weekly observations, the time support set has eight elements that correspond to the weeks during the construction period. In other words, the elements are week 1, week 2, . . ., week 8. The space support is used in relating the current state of a certain variable at a specific time support instant, to a specific location in space within the system. As an example, a space support set with elements that represent the type of structural element under construction is considered. These elements, for example, are columns, beams, slabs, and footings. Such a classification constitutes a space support set with four potential elements. The population support is used to represent the performance of units having the same structure with respect to the same variables. The population support set in this example can represent the set of the different crews involved in the construction activity. This support set, for example, has four potential elements, which

Chapter two: Information-based system definition

Table 2.2 States of Variables

Variable	States	Observation Channel o_i	Specific Variable v_i	Mapping Type	General Variable v'_i
Experience (v_1)	Poor	Maximum	2	One-to-one	0
	Fair	Maximum	5	One-to-one	1
	Good	Maximum	8	One-to-one	2
	Moderate	Maximum	9	One-to-one	3
	Excellent	Maximum	10	One-to-one	4
Method (v_2)	Traditional Method	One-to-one	Method 1	One-to-one	10
	Slip Form Method	One-to-one	Method 2	One-to-one	20
	Precast Method	One-to-one	Method 3	One-to-one	30

are falsework crew, rebar crew, concreting crew, and finishing crew. The overall support set, that represents the domain within which any of the defined variables can change, is defined by the Cartesian product of the three support sets. In other words, a certain variable is measured at a certain time instant in a certain location for a certain working crew. Therefore, the overall state of the attribute at a certain time instant is related to the performance and location of the working crew at that time. This fine classification allows for a complete identification of the reasons and factors that are responsible for a measured state of the attribute. This facilitates construction control process, and results in much more precise and accurate corrective actions. Table 2.2 summarizes different potential states for each variable together with observation channels (o_i) and corresponding general variables (v'_i). This example is based on the assumption that personnel with poor experience are not used in the construction activities. The observation channel is taken as a maximum operator to obtain the specific variable (v_i). For example, using the maximum operator on *poor* produces 2 from Figure 2.20. The mapping from (v_i) to (v'_i) is a one-to-one mapping that can be made for abstraction purposes to some generalized states. The tabulated values under (v'_i) in Table 2.2 were selected arbitrarily for demonstration purposes. Table 2.3 summarizes the different elements for each support set. Table 2.4 shows the overall support set for a combination of two of the supports considered in this example of time and space. For example, the pair [12, 11] in Table 2.4 indicates columns (i.e., general element 12, according to Table 2.3) and week 1 (i.e., general element 11, according to Table 2.3).

The source system defined as such is classified as neutral since an input/output identification was not considered. The defined variables are discrete. The time support set is linearly ordered while the space and population support sets are not ordered. Observation channels for variable (v_1) are linearly ordered, while these for variable (v_2) are not ordered. Observation

Table 2.3 Elements of the Different Support Sets

Support	Specific Element	Mapping Type	General Element
Time	week 1	One-to-one	11
	week 2	One-to-one	21
	week 3	One-to-one	31
	week 4	One-to-one	41
	week 5	One-to-one	51
	week 6	One-to-one	61
	week 7	One-to-one	71
	week 8	One-to-one	81
Space	Columns	One-to-one	12
	Beams	One-to-one	22
	Slabs	One-to-one	32
	Footings	One-to-one	42
Population	Falsework Crew	One-to-one	13
	Rebar Crew	One-to-one	23
	Concreting Crew	One-to-one	33
	Finishing Crew	One-to-one	43

Table 2.4 The Overall Support Set of Time and Space

Space	Time (Week)							
	Week (11)	Week (21)	Week (31)	Week (41)	Week (51)	Week (61)	Week (71)	Week (81)
Columns (12)	[12, 11]	[12, 21]	[12, 31]	[12, 41]	[12, 51]	[12, 61]	[12, 71]	[12, 81]
Beams (22)	[22, 11]	[22, 21]	[22, 31]	[22, 41]	[22, 51]	[22, 61]	[22, 71]	[22, 81]
Slabs (32)	[32, 11]	[32, 21]	[32, 31]	[32, 41]	[32, 51]	[32, 61]	[32, 71]	[32, 81]
Footings (42)	[42, 11]	[42, 21]	[42, 31]	[42, 41]	[42, 51]	[42, 61]	[42, 71]	[42, 81]

channels for variable (v_1) are fuzzy, while those for variable (v_2) are crisp. Figure 2.21 shows a block diagram of this source system.

2.3.2.2 Data systems

The second level of a hierarchical system classification is the data system, which includes a source system together with actual data introduced in the form of states of variables for each attribute. The actual states of the variables at the different support instances yield the overall states of the attributes. Special functions and techniques are used to infer information regarding an attribute, based on the states of the variables representing it. A formal definition of a data system could be expressed as follows:

$$D = \{S, a\} \qquad (2.2)$$

Chapter two: Information-based system definition 73

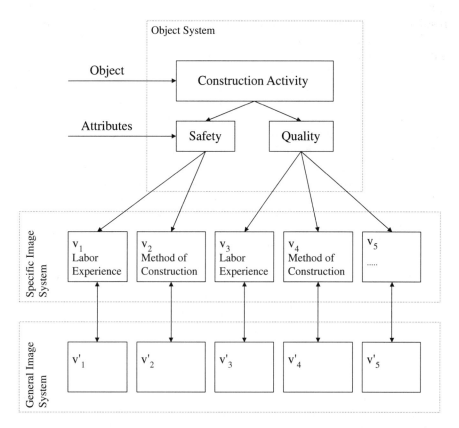

Figure 2.21 A source system of a construction activity.

where D = data system; S = the corresponding source system; and a = observed data that specify the actual states of the variables at different support instances.

Example 2.7   A data system for a construction activity

Considering the two variables previously defined, v_1 for labor experience and v_2 for method of construction, for the concrete placement construction activity of Example 2.6, example data are introduced to illustrate the formulation of the data system. Variable (v_1) was defined as a fuzzy variable with fuzzy observation channels. Accordingly, the data representing the states of this variable were also considered as fuzzy data. In other words, this variable can take its potential states at any support instant with certain degrees of belief. Considering the combination of time and space supports, this formulation results in a three-dimensional data matrix for variable (v_1). Any two-dimensional data matrix has the degrees of belief of each potential state as its entries. Variable (v_2) was defined as a crisp variable with crisp observation channels. As a result, the corresponding observed data were also crisp.

Considering the combination of time and space supports, this formulation results in a two-dimensional data matrix for variable (v_2) as its entries.

Data systems can be classified based on the level of available data. If all entries in a data matrix are specified, the system is known as completely specified. However, if some of the entries in a data matrix are not specified, the system is known as incompletely specified. Tables 2.5 and 2.6 show two examples for the two-dimensional data matrices representing two of the potential states of variable (v_1). Table 2.5 provides degrees of belief in having the state of good for v_1 as an example. Similar matrices are provided for other states as shown in Table 2.6. Table 2.7 shows a crisp data matrix for variable (v_2). Obviously in this example, all of the systems considered have completely specified data. Another classification or distinction that could be realized for data systems with linearly ordered support sets is periodic or nonperiodic data. Data are considered to be periodic if they repeat in the same order by extending the support set. From the data matrices specified in this example, such a property does not exist.

Table 2.5 The Data Matrix of Labor Experience (v_1) as Degrees of Belief in Having the State (GOOD)

Space	Time (Week)							
	11	21	31	41	51	61	71	81
12	0.7	0.5	0.6	0.1	0.3	0.2	0.8	1.0
22	1.0	0.4	0.7	0.5	0.7	1.0	0.9	0.3
32	0.2	0.7	1.0	0.9	0.3	0.5	1.0	0.6
42	0.9	0.5	0.8	0.7	0.5	0.2	0.1	0.3

Table 2.6 The Data Matrix of Labor Experience (v_1) as Degrees of Belief in Having the State (MODERATE)

Space	Time (Week)							
	11	21	31	41	51	61	71	81
12	0.3	0.7	0.9	1.0	0.5	0.3	0.2	0.8
22	0.9	0.5	0.7	0.6	1.0	0.9	0.5	0.6
32	0.3	0.9	1.0	0.8	0.2	0.7	0.9	1.0
42	0.3	0.5	0.7	1.0	0.6	0.8	0.4	0.2

Table 2.7 The Data Matrix of Method of Construction (v_2)

Space	Time (Week)							
	11	21	31	41	51	61	71	81
12	10	10	10	20	20	20	20	20
22	20	20	20	10	10	10	20	20
32	10	10	20	20	20	10	10	10
42	10	10	10	10	10	10	10	10

2.3.2.3 Generative systems

At the generative knowledge level, support independent relations are defined to describe the constraints among the variables. These relations could be utilized in generating states of the basic variables for a prescribed initial or boundary condition. The set of basic variables includes those defined by the source system and possibly some additional variables that are defined in terms of the basic variables. There are two main approaches for expressing these constraints. The first approach consists of a support independent function that describes the *behavior* of the system. A function defined as such is known as a *behavior function*. An example behavior function is provided at the end of the section in Example 2.8. The second approach consists of relating successive *states* of the different variables. In other words, this function describes a relationship between the current overall state of the basic variables and the next overall state of the same variables. A function defined as such is known as a *state-transition function*. An example state-transition function was provided in Example 2.3 using Markov chains. A generative system defined by a behavior function is referred to as a *behavior system*, whereas if it is defined by a state-transition function it is known as a *state-transition system*. State transition systems can always be converted into equivalent behavior systems, which makes the behavior systems more general.

The constraints among the variables at this level can be represented using many possible views or perspectives that are known as masks. A mask represents the pattern in the support set, that defines sampling variables, that should be considered. The sampling variables are related to the basic variables through translation rules that depend on the ordering of the support set. A formal definition of a behavior system could be expressed as

$$E_B = (I, K, f_B) \qquad (2.3)$$

where E_B = behavior system defined as triplet of three items; I = the corresponding general image system or the source system as a whole; K = the chosen mask; and f_B = behavior function. If the behavior function is used to generate data or states of the different variables, the sampling variables should be partitioned into generating and generated variables. The generating variables represent initial conditions for a specific generating scheme. The system in this form is referred to as a *generative behavior system*. The formal definition for such a system could be expressed as

$$E_{GB} = (I, K_G, f_{GB}) \qquad (2.4)$$

where E_{GB} = generative behavior system defined as triplet of three items; I = the corresponding general image system or the source system as a whole; K_G = the chosen mask partitioned into submasks, namely a generating submask which defines the generating sampling variables, and a generated submask which defines the generated variables; and f_{GB} = generative behavior function which should relate the occurrence of the general variables to that of the generating variables in a conditional format.

Most engineering and scientific models, such as the basic Newton's law of force computed as the product of mass of an object and its acceleration, or computing the stress in a rod under axial loading as the applied force divided by the cross sectional area of the rod, can be considered as generative systems that relate basic variables such as mass and acceleration to force, or axial force and area to stress, respectively. In these examples, these models are behavior systems.

Several methodological distinctions can be identified in this level. One of these distinctions is the type of behavior function used. For nondeterministic systems where variables have more than one potential state for the same support instant, a degree of belief or a probability measure to each potential state in the overall state set of the sampling variables should be assigned. This is accomplished by using any of the fuzzy measures. Figure 2.22 summarizes some of the different fuzzy measures and their interrelations based on Klir (1985). (These measures are discussed in detail in Chapter 5.) Each one of these measures is considered to form a certain distinction within the generative system. Probability distribution functions and possibility distribution functions are two of the most widely used behavior functions. The determination of a suitable behavior function for a given source system, mask, and data is not an easy task. Potential behavior functions should meet a set of conditions to be satisfactorily accepted. These conditions should be based on the actual constraints among the variables. They also relate to the degree of generative uncertainty and complexity of the behavior system. In other words, a selected behavior function should result in minimum disagreement due to constraints, minimum generative uncertainty, and minimum complexity.

Another distinction at this level could be identified in relation to the mask used. If the support set is ordered, the mask is known as *memory-dependent*; otherwise the mask is referred to as *memoryless*. Figure 2.23 summaries the different distinctions identified in this knowledge level. In the example under consideration, a probability distribution function is used as a behavior function. A memoryless mask is assumed that results in a set of sampling variables that are equivalent to the basic variables. All combinations of individual states of the different variables are considered. This process results in all potential overall states of the variables. A probability measure is then evaluated for each potential overall state. This measure depends on the frequency of occurrence of such an overall state and represents the value of the behavior function for that state. If crisp data are considered, the frequency of occurrence can be directly translated to the actual number of occurrences of such a state. However, if fuzzy data are considered, an appropriate aggregation function should be defined in order to evaluate the frequency of occurrence. For example, a multiplication function or a minimum function could form admissible candidates for such a

Chapter two: Information-based system definition

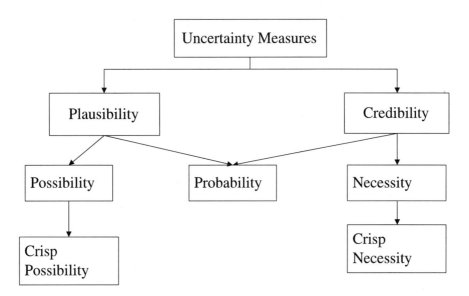

Figure 2.22 Uncertainty measures.

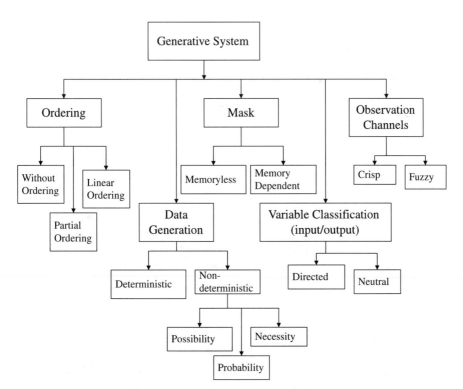

Figure 2.23 Methodological distinctions for generative systems.

task. These aggregation functions should be applied to the degrees of belief of the individual states of the different variables, which yield an overall degree of belief for each overall state. The sum of the degrees of belief of the different occurrences for the same overall state represents a measure for the likelihood of occurrence of that state. The corresponding probability of occurrence of each overall state is then calculated as the ratio of the likelihood of occurrence of such a state and the sum of all likelihoods of occurrence. In general, the problem of defining an acceptable behavior function starts with the determination of all the potential masks. A behavior function is then evaluated for each proposed mask as previously described. One behavior function should then be chosen based on the three properties mentioned earlier, namely minimum disagreement, minimum generative uncertainty, and minimum complexity.

Example 2.8 A generative system for a construction activity

A memoryless mask was chosen in this example for illustration purposes. In Table 2.8, labor experience variable (v_1) was defined as a fuzzy variable that can take state (1) at different support instances with the degrees of belief shown in the table. This state was accompanied by state (10) for construction method variable (v_2) as shown in the table. Accordingly, the overall state (C_1) = (1, 10) has support-variant degrees of belief. Using a minimum operator, for example, as an aggregation function, the degree of belief of state (C_1) can be calculated at the different support instants as shown in Table 2.8. In other words, the degree of belief of the combination of states (1) and (10) is the minimum of the two degrees of belief of the separate states. It should be noted that since variable (v_2) is a crisp variable, its degree of belief was taken to be one at any support instant. The likelihood of occurrence of each overall state (C_1) was then calculated as follows:

$$N_c = \sum_{\text{all t}} d_{s,t} \tag{2.5}$$

where N_c = likelihood of occurrence; $d_{s,t}$ = aggregated degree of belief of state(s) at support instant (t); and the summation was performed over the support instances. The corresponding probability of overall state (C_1) is then calculated using the following formula (Klir, 1969 and 1985):

$$f_B(C_1) = \frac{N_c}{\sum_{\text{all c}} N_c} \tag{2.6}$$

where $f_B(C_1)$ = probability of having state (C_1) which corresponds to the value of the behavior function for that state; N_c = likelihood of occurrence of

Table 2.8 A Behavior Function Evaluation for Variables (v_1) and (v_2)

Overall State (C_i)	Variable (v_1)		Variable (v_2)		Degree of Belief of Overall State (C)	Likelihood of Occurrence (N_c)	Behavior Function (f_B)
	State	Degree of Belief	State	Degree of Belief			
C_1 (1, 10)	1	0.8	10	1	0.8		
	1	0.7	10	1	0.7		
	1	0.5	10	1	0.5		
	1	0.3	10	1	0.3	2.3	0.354
C_2 (3, 10)	3	0.4	10	1	0.4		
	3	0.7	10	1	0.6		
	3	0.6	10	1	0.7	1.7	0.262
C_3 (2, 10)	2	0.5	10	1	0.5		
	2	0.8	10	1	0.8		
	2	0.9	10	1	0.9	2.2	0.338
C_4 (0, 10)	0	0.2	10	1	0.2		
	0	0.1	10	1	0.1	0.3	0.046

state (C_1); and the summation was performed over all the overall states. There are other expressions defined in the literature for the calculation of probabilities or possibilities based on fuzzy and crisp data (Ayyub and McCuen, 1998; Ang and Tang, 1975; Klir, 1985). The expressions provided by Equations 2.4 and 2.5 were chosen for illustration purposes. The resulting probabilities (f_B) for selected potential overall states are shown in Table 2.8.

A state transition system can be expressed as

$$E_n = (I, K, f_n) \tag{2.7}$$

where E_n = a state-transition system; I = the corresponding general image system; K = the chosen mask; and f_n = the state-transition function. An important interpretation of the state transition concept in construction is the state table approach as used by Abraham et al. (1989). The state table format could be viewed as a *state-transition function* in a feedback control framework. Table 2.9 shows an example of such a table that describes the process of giving some command, the current state of a certain variable, the next state for the same variable, and a feedback information for control purposes. The main concept in this framework is the relationship developed through the table, between the consecutive states of the different variables. This support-independent relationship between the successive states of the different variables represents a state-transition function. Figure 2.24 shows a block diagram of the defined generative system. It should be emphasized here that although variables 1 and 2 for attribute 1 have the same names as variables 3 and 4 for attribute 2, they would not necessarily take the same values. In other words, the same variables have different impacts on different attributes according to the nature of each attribute.

Table 2.9 A State Table Format (Abraham 1989)

Command	State	Feedback	Next State	Output	Report
Place concrete for a foundation	Forms without concrete	Concrete overflow	Structural member	Concrete member	Concrete quantities
Place concrete for a column	Forms without concrete	Concrete overflow	Structural member	Concrete member	Concrete quantities
Place concrete for a beam	Forms without concrete	Concrete overflow	Structural member	Concrete member	Concrete quantities
Place concrete for a slab	Forms without concrete	Concrete overflow	Structural member	Concrete member	Concrete quantities

Chapter two: Information-based system definition 81

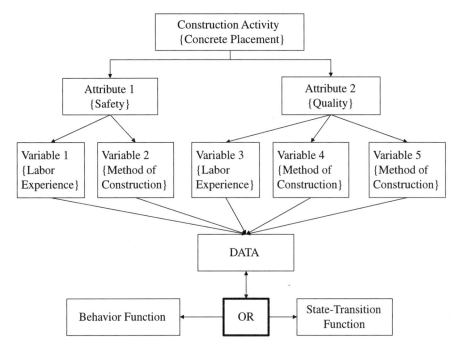

Figure 2.24 A generative system for a construction activity.

2.3.2.4 Structure systems

Structure systems are sets of smaller systems or subsystems. The subsystems could be source, data, or generative systems. These subsystems may be coupled due to having common variables or due to interaction in some other form. A formal definition of a structure system could be expressed as follows:

$$SE_B = \{(V_i, E_B^i), \text{ for all } i \in e\} \tag{2.8}$$

where SE_B = structure system whose elements are behavior systems; V_i = the set of sampling variables for the element of the behavior system; E_B^i = i^{th} behavior system; and e = the total number of elements or subsystems in the structure system with all i that belong to e, i.e., for all $i \in e$.

Example 2.9   A structure system for a construction activity

In the construction example, the construction activity is viewed as a structure system. The construction activity consists of a number of processes that should be accomplished. These processes depend on each other in some manner. Considering concrete placement as a construction activity, the different processes involved include falsework construction, rebar placement, concrete pouring, and concrete finishing. These processes represent interre-

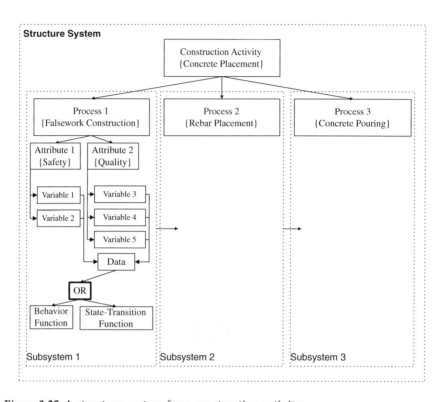

Figure 2.25 A structure system for a construction activity.

lated subsystems within the structure system. Each process is defined as a generative system. The interrelation among the subsystems represents the dependence of each process on the preceding one. Another form of the interrelationship is the input/output relation between the successive processes. A nested structure system could also be defined on the same example by defining each of the subsystems as another structure system whose elements are generative systems. It should be noted that each of the described subsystems and their corresponding elements should be defined on the same source system. Figure 2.25 shows a block diagram for the structure system of the construction activity.

2.3.2.5 Metasystems

Metasystems are introduced for the purpose of describing changes within a given support set. The metasystem consists of a set of systems defined at some lower knowledge level and some support-independent relation. Referred to as a replacement procedure, this relation defines the changes in the lower level systems. All the lower level systems should share the same source system. There are two different approaches whereby a metasystem could be viewed in relation to the structure system. The first approach is introduced by defining the system as a structure metasystem. Considering the construction example, the construction activity is defined as a structure system whose elements are metasystems. Each metasystem represents the change in its behavior system. As an example for a concrete placement activity, the processes include falsework construction, rebar placement, and concrete pouring. However, in order to represent the actual behavior of this system within the overall support set required, the behavior system in this case can only be defined using more than one behavior function. Each one of these functions is valid for only a subset of the overall support set. Stating the problem in this format, a metasystem should be defined to describe the change in each one of these subsystems. The replacement procedure is required in order to decide which behavior function should be used. This decision should be taken based on the states of some basic variables specified for this purpose. Referring to the behavior functions previously defined, i.e., the probability/possibility distributions, more than one distribution might be necessary to fit the available date. Each one of these distributions is valid within a subset of the overall support set. For example, some distribution might fit variable (v_1) for the first month, i.e., four weeks, while a different distribution might more appropriately represent the same variable during the next four weeks. Thus, a replacement procedure is required in order to calculate the current time and choose the appropriate distribution that represents the data during this period. Figure 2.26 shows a graphical representation of a metasystem.

The second approach consists of defining a metasystem of a structure system whose elements are behavior systems. Applying this concept to the example under consideration, a metasystem is defined on the construction activity that represents a structure system. In other words, the construction activity is the structure system with the different processes defined as subsystems. This structure system changes with time, where time is as an example support. At some instant of time, falsework construction and rebar replacement as two subsystems might be in progress, whereas the other processes, i.e., concrete pouring and concrete finishing, might not be started yet. Therefore, the components of the structure system at this instant of time do not include the ones that have not been started. This composition would probably be different after some time, where all the processes might be in progress at the same time. The replacement procedure in this case should observe the change in each process such that the composition of the structure system could be defined at any support instant.

2.4 Models for ignorance and uncertainty types

2.4.1 Mathematical theories for ignorance types

Systems analysis provides a general framework for modeling and solving various problems and making appropriate decisions. For example, an engineering model of an engineering project starts by defining the system includ-

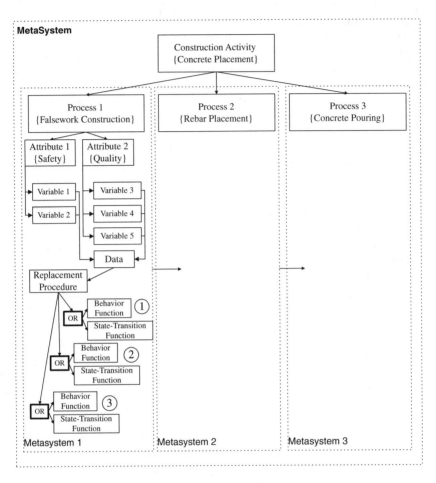

Figure 2.26 A metasystem for a construction activity.

ing a segment of the project's environment that interacts significantly with it. The limits of the system are drawn based on the nature of the project, class of performances (including failures) under consideration, and the objectives of the analysis. The system definition can be based on observations at different system levels in the form of a hierarchy. The observations can be about the source and data elements, interactions among these elements, and behavior of the system as was discussed in Section 2.3. Each level of knowledge that is obtained about an engineering problem can be said to define a system on the problem. As additional levels of knowledge are added to previous ones, higher epistemological levels of system definition and description are generated which, taken together, form a hierarchy of such system descriptions. An epistemological hierarchy of systems suited to the representation of engineering problems with a generalized treatment of uncertainty can provide realistic assessments of systems (Klir, 1985; Klir and Folger, 1988).

Chapter 1 deals with knowledge and ignorance, and their nature, types, and sources as summarized in Figures 1.4 and 1.9, respectively. The nature, types, and sources of knowledge were examined by philosophers, scientists, and engineers as provided in Chapter 1, whereas ignorance has received less attention. However, interest in analyzing and modeling uncertainty and ignorance was started by the works of Zadeh (1965 and 1978), Dempster (1976a and 1976b), Shafer (1976), Sugeno (1974 and 1977), Klir and Folger (1988), Pawlak (1991), and Smithson (1989) by suggesting various models and theories for representing ignorance categories that are of the conscious ignorance types. Table 2.10 maps various theories that are suitable for ignorance modeling; however, in solving problems in engineering and science that involve several ignorance types, combinations of these theories are needed. Each problem might require a mix of theories that most appropriately and effectively models its ignorance content.

According to Table 2.10, classical sets theory can effectively deal with ambiguity by modeling nonspecificity, whereas fuzzy and rough sets can be used to model vagueness, coarseness, and simplifications. The theories of probability and statistics are commonly used to model randomness and sampling uncertainty. Bayesian methods can be used to combine randomness or sampling uncertainty with subjective information that can be viewed as a form of simplification. Ambiguity, as an ignorance type, forms a basis for randomness and sampling, hence its cross-shading in the table with classical sets, probability, statistics, and Bayesian methods. Inaccuracy, as an ignorance type that can be present in many problems, is cross-shaded in the table with probability, statistics, and Bayesian methods. The theories of evidence, possibility, and monotone measure can be used to model confusion, conflict, and vagueness. Interval analysis can be used to model vagueness and simplification, whereas interval probabilities can be used to model randomness and simplification.

2.4.2 *Information uncertainty in engineering systems*

2.4.2.1 *Abstraction and modeling of engineering systems*

Uncertainty modeling and analysis in engineering started with the employment of safety factors using deterministic analysis, then was followed by probabilistic analysis with reliability-based safety factors. Uncertainty in engineering was traditionally classified into objective and subjective types. The objective types included the physical, statistical, and modeling sources of uncertainty. The subjective types were based on lack of knowledge and expert-based assessment of engineering variables and parameters. This classification was still deficient in completely covering the entire nature of uncertainty. The difficulty in completely modeling and analyzing uncertainty stems from its complex nature and its invasion of almost all epistemological levels of a system by varying degrees, as discussed in Chapter 1.

Table 2.10 Theories to Model and Analyze Ignorance Types

Theory	Confusion & Conflict	Inaccuracy	Ambiguity	Randomness & Sampling	Vagueness	Coarseness	Simplification
Classical sets	■						
Probability		■	■	■			
Statistics		■	■	■			
Bayesian				■			
Fuzzy sets					■		■
Rough sets	■					■	■
Evidence	■				■		
Possibility					■		
Monotone measure				■			
Interval probabilities							■
Interval analysis							■

Engineers can deal with information for the purpose of system analysis and design. Information in this case is classified, sorted, analyzed, and used to predict system attributes, variables, parameters, and performances. However, it can be more difficult to classify, sort, and analyze the uncertainty in this information and use it to assess uncertainties in our predictions.

Uncertainties in engineering systems can be mainly attributed to ambiguity, likelihood, approximations, and inconsistency in defining the architecture, variables, parameters and governing prediction models for the systems. The ambiguity component comes from either not fully identifying possible outcomes or incorrectly identifying possible outcomes. Likelihood builds on the ambiguity of defining all the possible outcomes by introducing probabilities to represent randomness and sampling. Therefore, likelihood includes the sources (1) physical randomness and (2) statistical uncertainty due to the use of sampled information to estimate the characteristics of the population parameters. Simplifications and assumptions, as components of approximations, are common in engineering due to the lack of knowledge and the use of analytical and prediction models, simplified methods, and idealized representations of real performances. Approximations also include vagueness and coarseness. The vagueness-related uncertainty is due to sources that include (1) the definition of some parameters, e.g., structural performance (failure or survival), quality, deterioration, skill and experience of construction workers and engineers, environmental impact of projects, conditions of existing structures using linguistic measures; (2) human factors; and (3) defining the interrelationships among the parameters of the problems, especially for complex systems. The coarseness uncertainty can be noted in simplification models and behavior of systems. Other sources of ignorance include inconsistency with its components of conflict and confusion of information and inaccuracies due to, for example, human and organizational errors.

Analysis of engineering systems commonly starts with a definition of a system that can be viewed as an abstraction of the real system. The abstraction is performed at different epistemological levels (Ayyub, 1992 and 1994). The process of abstraction can be graphically represented as shown in Figure 2.27. A resulting model from this abstraction depends largely on the engineer (or analyst) who performed the abstraction, hence on the subjective nature of this process. During the process of abstraction, the engineer needs to make decisions regarding what aspects should or should not be included in the model. These aspects are shown in the Figure 2.27. Aspects that are abstracted and not abstracted include the previously identified uncertainty types. In addition to the abstracted and nonabstracted aspects, unknown aspects of the system can exist due to blind ignorance, and they are more difficult to deal with because of their unknown nature, sources, extents, and impact on the system.

In engineering, it is common to perform uncertainty modeling and analysis of the abstracted aspects of the system with a proper consideration of the nonabstracted aspects of a system. The division between abstracted and

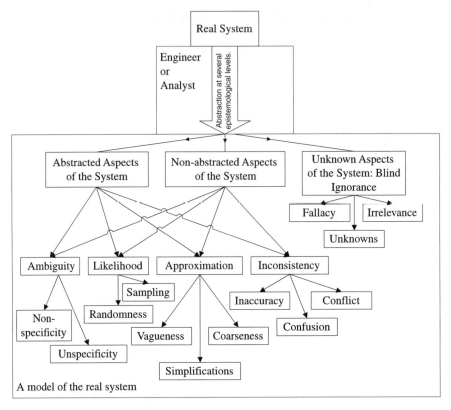

Figure 2.27 Abstraction and ignorance for engineering systems.

nonabstracted aspects can be a division of convenience that is driven by the objectives of the system modeling, or simplification of the model. However, the unknown aspects of the systems are due to blind ignorance that depends on the knowledge of the analyst and the state of knowledge about the system in general. The effects of the unknown aspects on the ability of the system model to predict the behavior of the real system can range from none to significant.

2.4.2.2 Ignorance and uncertainty in abstracted aspects of a system

Engineers and researchers dealt with the ambiguity and likelihood types of uncertainty in predicting the behavior and designing engineering systems using the theories of probability and statistics and Bayesian methods. Probability distributions were used to model system parameters that are uncertain. Probabilistic methods that include reliability methods, probabilistic engineering mechanics, stochastic finite element methods, reliability-based design formats, and other methods were developed and used for this purpose. In this treatment, however, a realization was established of the presence of the approximations type of uncertainty. Subjective probabilities were used to deal with this type that are based on mathematics used for the frequency-type of probability. Uniform and triangular probability distributions were used to model this type of uncertainty for some parameters. The Bayesian techniques were also used, for example, to deal with combining empirical and subjective information about these parameters. The underlying distributions and probabilities were, therefore, updated. Regardless of the nature of uncertainty in the gained information, similar mathematical assumptions and tools were used that are based on probability theory.

Approximations arise from human cognition and intelligence. They result in uncertainty in mind-based abstractions of reality. These abstractions are, therefore, subjective and can lack crispness, they can be coarse in nature, or they might be based on simplifications. The lack of crispness, called vagueness, is distinct from ambiguity and likelihood in source and natural properties. The axioms of probability and statistics are limiting for the proper modeling and analysis of this uncertainty type and are not completely relevant, nor completely applicable. The vagueness type of uncertainty in engineering systems can be dealt with using appropriately fuzzy set theory (Zadeh, 1965). Fuzzy set theory has been developed by Zadeh (1965, 1968, 1973, 1975, 1978) and used by scientists, researchers, and engineers in many fields. Example applications are provided elsewhere (Kaufmann and Gupta, 1985; Kaufmann, 1975). In engineering, the theory was proven to be a useful tool in solving problems that involve the vagueness type of uncertainty. For example, civil engineers and researchers started using fuzzy sets and systems in the early 1970s (Brown, 1979 and 1980; Brown and Yao, 1983). To date, many applications of the theory in engineering were developed. The theory has been successfully used in, for example (Ayyub, 1991; Blockley, 1975, 1979a, 1979b, 1980; Blockley et al., 1983; Shiraishi and Furuta, 1983; Shiraishi et al., 1985; Yao, 1979, 1980; Yao and Furuta, 1986; Blockley et al., 1975 to 1983; Furuta et al., 1985 and 1986; Ishizuka et al., 1981 and 1983; Itoh and Itagaki, 1989; Kaneyoshi, 1990; Shiraishi et al., 1983 and 1985; Yao et al. 1979, 1980, 1986),

- strength assessment of existing structures and other structural engineering applications;
- risk analysis and assessment in engineering;
- analysis of construction failures, scheduling of construction activities, safety assessment of construction activities, decisions during construction and tender evaluation;
- the impact assessment of engineering projects on the quality of wildlife habitat;
- planning of river basins;
- control of engineering systems;
- computer vision; and
- optimization based on soft constraints.

Coarseness in information can arise from approximating an unknown relationship or set by partitioning the universal space with associated belief levels for the partitioning subsets in representing the unknown relationship or set (Pawlak, 1991). Such an approximation is based on *rough sets* as described in Chapter 4. Pal and Skowron (1999) provide background and detailed information on rough set theory, its applications, and hybrid fuzzy-rough set modeling. Simplifying assumptions are common in developing engineering models. Errors resulting from these simplifications are commonly dealt with in engineering using bias random variables that are assessed empirically. A system can also be simplified by using knowledge-based if-then rules to represent its behavior based on fuzzy logic and approximate reasoning.

2.4.2.3 Ignorance and uncertainty in nonabstracted aspects of a system

In developing a model, an analyst or engineer needs to decide, at the different levels of modeling a system, upon the aspects of the system that need to be abstracted and the aspects that need not to be abstracted. The division between abstracted and nonabstracted aspects can be for convenience or to simplify the model. The resulting division can depend on the analyst or engineer, as a result of his or her background, and the general state of knowledge about the system.

The abstracted aspects of a system and their uncertainty models can be developed to account for the nonabstracted aspects of the system to some extent. Generally, this accounting process is incomplete. Therefore, a source of uncertainty exists due to the nonabstracted aspects of the system. The ignorance categories and uncertainty types in this case are similar to the previous case of abstracted aspects of the system. These categories and types are shown in Figure 2.27.

The ignorance categories and uncertainty types due to the nonabstracted aspects of a system are more difficult to deal with than the uncertainty types due to the abstracted aspects of the system. The difficulty can stem from a lack of knowledge or understanding of the effects of the nonabstracted aspects on the resulting model in terms of its ability to mimic the real system. Poor judgment or human errors about the importance of the nonabstracted aspects of the system can partly contribute to these uncertainty types, in addition to contributing to the next category, uncertainty due to the unknown aspects of a system.

2.4.2.4 Ignorance due to unknown aspects of a system

Some engineering failures have occurred because of failure modes that were not accounted for in the design stages of these systems. The nonaccounting for the failure modes can be due to (1) blind ignorance, negligence, using irrelevant information or knowledge, human errors, or organizational errors; or (2) a general state of knowledge about a system that is incomplete. These unknown system aspects depend on the nature of the system under consideration, the knowledge of the analyst, and the state of knowledge about the system in general. The nonaccounting of these aspects in the models for the system can result in varying levels of impact on the ability of these models in mimicking the behavior of the systems. The effects of the unknown aspects on these models can range from none to significant. In this case, the ignorance categories include wrong information and fallacy, irrelevant information, and unknowns as shown in Figure 2.27.

Engineers dealt with nonabstracted and unknown aspects of a system by assessing what is commonly called the modeling uncertainty, defined as the ratio of a predicted system's variables or parameters (based on the model) to the value of the parameter in the real system. This empirical ratio, which is called the bias, is commonly treated as a random variable that can consist of objective and subjective components. Factors of safety are intended to safeguard against failures. This approach of bias assessment is based on two implicit assumptions: (1) the value of the variable or parameter for the real system is known or can be accurately assessed from historical information or expert judgment; and (2) the state of knowledge about the real system is complete or bounded, and reliable. For some systems, the first assumption can be approximately examined through verification and validation, whereas the second assumption generally cannot be validated.

2.5 System complexity

Our most troubling long-range problems, such as economic forecasting and trade balance, defense systems, and genetic modeling, center on systems of extraordinary complexity. The systems that host these problems — computer networks, economics, ecologies, and immune systems — appear to be as diverse as the problems. Humans as complex, intelligent systems have the ability to anticipate the future, learn, and adapt in ways that are not yet fully understood. Engineers and scientists, who study or design systems, have to deal with complexity more often than ever, hence the interest in the field of complexity. Understanding and modeling system complexity can be viewed as a pretext for solving complex scientific and technological problems, such as finding a cure for the acquired immune deficiency syndrome (AIDS), solving long-term environmental issues, or using genetic engineering safely in agricultural products. The study of complexity led to, for example, chaos and catastrophe theories. Even if complexity theories would not produce solutions to problems, they can still help us to understand complex systems and perhaps direct experimental studies. Theory and experiment go hand in glove, therefore providing opportunities to make major contributions.

The science of complexity was founded at the Santa Fe Institute by a group of physicists, economists, mathematicians, and computer scientists that included Nobel Laureates in physics and economics, Murray Gell-Mann and Kenneth Arrow, respectively. They noted that scientific modeling and discovery tends to emphasize linearity and reductionism, and they consequently developed the science of complexity based on assumed interconnectivity, co-evolution, chaos, structure, and order to model nature, human social behavior, life, and the universe in unified manners (Waldrop 1992).

Complexity can be classified into two broad categories: (1) complexity with structure, and (2) complexity without structure. The complexity with structure was termed *organized complexity* by Weaver (1948). Organized complexity can be observed in a system that involves nonlinear differential equations with many interactions among a large number of components and variables that define the system, such as in life, behavioral, social, and environmental sciences. Such systems are usually nondeterministic in their nature. Problem solutions related to such models of organized complexity tend to converge to statistically meaningful averages (Klir and Wierman, 1999). Advancements in computer technology and numerical methods have enhanced our ability to obtain such solutions effectively and inexpensively. As a result, engineers design complex systems in simulated environments and operations, such as a space mission to a distant planet, and scientists can conduct numerical experiments involving, for example, nuclear blasts.

In the area of simulation-based design, engineers are using parallel computing and physics-based modeling to simulate fire propagation in engineering systems, or the turbulent flow of a jet engine using molecular motion and modeling. These computer and numerical advancements are not limitless, as the increasing computational requirements lead to what is termed *transcomputational problems* capped by the *Bremermann's limit* (Bremermann, 1962). The nature of such transcomputational problems is studied by the theory of computational complexity (Garey and Johnson, 1979). The Bremermann's limit was estimated based on quantum theory using the following proposition (Bremermann, 1962):

> "No data processing systems, whether artificial or living, can process more than 2×10^{47} bits per second per gram of its mass,"

where data processing is defined as transmitting bits over one or several of a system's channels. Klir and Folger (1988) provide additional information on the theoretical basis for this proposition, showing that the maximum processing value is to be 1.36×10^{47} bits per second per gram of its mass. Considering a hypothetical computer that has the entire mass of the Earth operating for a time period equal to an estimated age of the Earth, i.e., 6×10^{27} grams and 10^{10} years, respectively, with each year containing 3.15×10^{7} seconds, this imaginary computer would be able to process 2.57×10^{92} bits, or rounded to the nearest power of ten, 10^{93} bits, defining the Bremermann's limit. Many scientific and engineering problems defined with many details can exceed this limit. Klir and Folger (1988) provide the examples of pattern recognition and human vision that can easily reach transcomputational levels. In pattern recognition, consider a square $q \times q$ spatial array defining

Chapter two: Information-based system definition

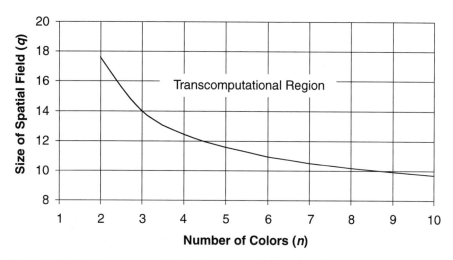

Figure 2.28 The Bremermann's limit for pattern recognition.

$n = q^2$ cells that partition the recognition space. Pattern recognition often involves color. Using k colors, as an example, the number of possible color patterns within the space is k^n. In order to stay within the Bremermann's limit, the following inequality must be met:

$$k^n \leq 10^{93} \quad (2.9)$$

Figure 2.28 shows a plot of this inequality for values of $k = 2$ to 10 colors. For example using only two colors, a transcomputational state is reached at $q \geq 18$ colors. These computations in pattern recognition can be directly related to human vision and the complexity associated with processing information by the retina of a human eye. According to Klir and Folger (1988), if we consider a retina of about one million cells with each cell having only two states, *active* and *inactive*, in recognizing an object, modeling the retina in its entirety would require the processing of

$$2^{1,000,000} = 10^{300} \quad (2.10)$$

bits of information, far beyond the Bremermann's limit.

Generally, an engineering system needs to be modeled with a portion of its environment that interacts significantly with it in order to assess some system attributes of interest. The level of interaction with the environment can be assessed only subjectively. By increasing the size of the environment and level of details in a model of the system, the complexity of the system model increases, possibly in a manner that does not have a recognizable or observable structure. This complexity without structure is more difficult to model and deal with in engineering and sciences. By increasing the complexity of the system model, our ability to make relevant assessments of the system's attributes can diminish. Therefore, there is a tradeoff between relevance and precision in system modeling in this case. Our goal should be to model a system with a sufficient level of detail that can result in sufficient precision and can lead to relevant decisions in order to meet the objective of the system assessment.

Living systems show signs of these tradeoffs between precision and relevance in order to deal with complexity. The survival instincts of living systems have evolved and manifest themselves as processes to cope with complexity and information overload. The ability of a living system to make relevant assessments diminishes with increased information input, as discussed by Miller (1978). Living systems commonly need to process information in a continuous manner in order to survive. For example, a fish needs to process visual information constantly in order to avoid being eaten by another fish. When a school of larger fish rushes towards the fish, presenting it with images of threats and dangers, the fish might not be able to process all the information and images become confused. Considering the information processing capabilities of living systems as input-output black boxes, the input and output to such systems can be measured and plotted in order to examine such relationships and any nonlinear characteristics that they might exhibit. Miller (1978) described these relationships for living systems using the following hypothesis that was analytically modeled and experimentally validated:

> As the information input to a single channel of a living system — measured in bits per second — increases, the information output — measured similarly — increases almost identically at first but gradually falls behind as it approaches a certain output rate, the channel capacity, which cannot be exceeded. The output then levels off at that rate, and finally, as the information input rate continues to go up, the output decreases gradually towards zero as breakdown or the confusion state occurs under overload.

The above hypothesis was used to construct families of curves to represent the effects of information input overload, as shown schematically in Figure 2.29. Once the input overload is removed, most living systems recover instantly from the overload, and the process is completely reversible; however, if the energy level of the input is much larger than the channel capacity, a living system might not fully recover from this input overload. Living systems also adjust the way they process information in order to deal with an information input overload using one or more of the following processes by varying degrees depending on the level of a living system in terms of complexity: (1) omission by failing to transmit information, (2) error by transmitting information incorrectly, (3) queuing by delaying transmission, (4) filtering by giving priority in processing, (5) abstracting by processing

Chapter two: Information-based system definition

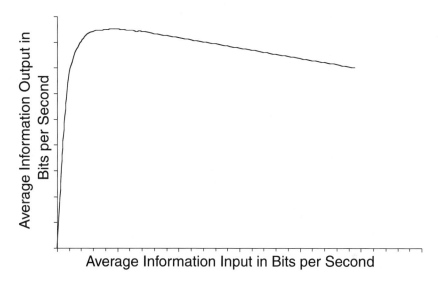

Figure 2.29 A schematic relationship of input and output information transmission rates for living systems.

messages with less than complete details, (6) multiple channel processing by simultaneously transmitting messages over several parallel channels, (7) escape by acting to cut off information input, and (8) chunking by transforming information into meaningful chunks. These actions can also be viewed as simplification means to cope with complexity and/or an information input overload.

2.6 Exercise problems

Problem 2.1. Provide example performance and functional requirements for an office building. Develop portions of a work breakdown structure for the office building.

Problem 2.2. Provide example performance and functional requirements for a residential house. Develop portions of a work breakdown structure for the house.

Problem 2.3. Develop and discuss a system engineering process for a low-income townhouse as an engineering product.

Problem 2.4. Develop and discuss the lifecycle of a major highway bridge as an engineering system.

Problem 2.5. Describe three engineering systems that can be modeled using the black-box method. What are the inputs and outputs for each system?

Problem 2.6. Describe three natural systems that can be modeled using the black-box method. What are the inputs and outputs for each system?

Problem 2.7. Describe three engineering systems that can be modeled using the state-based method. What are the states for each system?

Problem 2.8. Describe three natural systems that can be modeled using the state-based method. What are the states for each system?

Problem 2.9. Create a structure for a decision problem in engineering similar to Figure 2.14.

Problem 2.10. Create a structure for a decision problem in investment similar to Figure 2.14.

Problem 2.11. Build an information-based hierarchical system definition for an office building by defining the source system, data system, generative system, structure system, and metasystem.

Problem 2.12. Repeat problem 2.11 for a highway bridge.

Problem 2.13. Repeat problem 2.11 for a residential house.

Problem 2.14. Provide engineering examples of structured and unstructured complexity.

Problem 2.15. Provide examples in science of structured and unstructured complexity.

Problem 2.16. Provide two cases of transcomputational problems. Why are they transcomputational in nature?

chapter three

Experts, opinions, and elicitation methods

Contents

3.1. Introduction 97
3.2. Experts and expert opinions 98
3.3. Historical background 99
     3.3.1. Delphi method 99
     3.3.2. Scenario analysis 105
3.4. Scientific heuristics 105
3.5. Rational consensus 109
3.6. Elicitation methods 111
     3.6.1. Indirect elicitation 111
     3.6.2. Direct method 113
     3.6.3. Parametric estimation 114
3.7. Standards for educational and psychological testing 114
3.8. Methods of social research 118
3.9. Focus groups 123
3.10. Exercise problems 124

3.1 Introduction

Decision and policy makers are routinely interested in speculative knowledge using experts for their opinions. Knowledge categories based on these sources were identified as *pistis* and *eikasia* in Figure 1.5. These categories are not new, but creating a structured mechanism for their acquisition and elicitation started relatively recently, after World War II. Expert-opinion elicitation reachesd its peak in terms of public confidence in its results during the Vietnam War, followed by a decline towards the end of the Nixon administration (Cooke, 1991). However, there is a renewed interest in these methods that can be attributed in part to the needs for technology forecasting and reliability and risk studies for dealing with technology, environment, and socioeconomic problems and challenges.

The objective of this chapter is to provide background information on experts, opinions, expert-opinion elicitation methods, methods used in developing questionnaires in educational and psychological testing and social research, and methods and practices utilized in focus groups.

3.2 Experts and expert opinions

An *expert* can be defined as a very skillful person who had much training and has knowledge in some special field. The expert is the provider of an opinion in the process of expert-opinion elicitation. Someone can become an expert in some special field by having the training and knowledge to a publicized level that would make him or her recognized by others as such. Figure 1.7 shows an expert (A) of some knowledge about the system represented using ellipses for illustrative purposes. Three types of ellipses were identified: (1) a subset of the evolutionary infallible knowledge (EIK) that the expert has learned, captured and/or created, (2) self-perceived knowledge by the expert, and (3) perception by others of the expert's knowledge. As was noted in Chapter 1, the EIK of the expert might be smaller than the self-perceived knowledge by the expert, and the difference between the two types is a measure of overconfidence that can be partially related to the expert's ego. Ideally, the three ellipses should be the same, but commonly they are not since they are greatly affected by communication skills of experts. Figure 1.7 also shows how the expert's knowledge can extend beyond the reliable knowledge base into the EIK area as a result of creativity and imagination of the expert. Another expert (i.e., Expert B) would have her/his own ellipses that might overlap with the ellipses of Expert A and might overlap with other regions by varying magnitudes. It can be noted from Figure 1.7 that experts might unintentionally provide opinions that are false.

An *expert opinion* can be defined as the formal judgment of an expert on a matter, also called an issue, in which his or her advice is sought. Also, an opinion could mean a judgment or a belief that is based on uncertain information or knowledge. An opinion is a subjective assessment, evaluation, impression, or estimation of the quality or quantity of something of interest that seems true, valid, or probable to the expert's own mind. In the legal practice, an opinion means an inconclusive judgment, which while it remains open to dispute seems true or probable to an expert's own mind. The Webster's dictionary also provides the following meanings to an opinion: (1) as a belief which means the mental acceptance of an idea or conclusion, often a doctrine or dogma proposed to one for acceptance; (2) as a view which is an opinion affected by one's personal manner of looking at things; (3) as a conviction which is a strong belief about whose truth one has no doubts; (4) as sentiment which refers to an opinion that is the result of deliberation but is colored with emotion; and (5) as persuasion which refers to a strong belief that is unshakable because one wishes to believe in its truth.

3.3 Historical background

The development of structured methods for expert-opinion elicitation was done by the RAND (**R**esearch **AND** **D**evelopment) Corporation of Santa Monica, California. The RAND Corporation resulted from a joint U.S. Air Force and Douglas Aircraft effort in 1946 called *Project RAND*. In its first year of operation, RAND predicted the first space satellite would be launched in the middle of 1957. The prediction was accurately validated by the Russian Sputnik launch on October 4, 1957. In 1948, RAND split from Douglas Aircraft as the first think-tank type of a corporation. The research of RAND was classified into four broad categories: (1) methodology, (2) strategic and tactical planning, (3) international relations, and (4) new technology. Almost all of these categories can rely heavily on expert opinions. In its early days between World War II and the Vietnam War, RAND developed two methods for structured elicitation of expert opinions: (1) Delphi method, and (2) scenario analysis.

3.3.1 Delphi method

The Delphi method is by far the most known method for eliciting and synthesizing expert opinions. RAND developed the Delphi method for the U.S. Air Force in the 1950s. In 1963, Helmer and Gordon used the Delphi method for a highly publicized long-range forecasting study on technological innovations (Helmer, 1968). The method was extensively used in a wide variety of applications in the 1960s and 1970s, exceeding 10,000 studies in 1974 on primarily technology forecasting and policy analysis (Linstone and Turoff, 1975).

The purpose and steps of the Delphi method depend on the nature of use. Primarily the uses can be categorized into technological forecasting and policy analysis. The technological forecasting relies on a group of experts on a subject matter of interest. The experts should be the most knowledgeable about issues or questions of concern. The issues and/or questions need to be stated by the study facilitators or analysts or by a monitoring team, and a high degree of consensus is sought from the experts. On the other hand, the policy analysis Delphi method seeks to incorporate the opinions and views of the entire spectrum of stakeholders and seeks to communicate the spread of opinions to decision-makers. For the purposes of this book, we are generally interested in the former type of consensus opinion.

The basic Delphi method consists of the following steps (Helmer, 1968):

1. Selection of issues or questions and development of questionnaires.
2. Selection of experts who are most knowledgeable about issues or questions of concern.
3. Issue familiarization of experts by providing sufficient details on the issues on the questionnaires.
4. Elicitation of experts about the issues. The experts might not know who the other respondents are.

5. Aggregation and presentation of results in the form of median values and an interquartile range (i.e., 25% and 75% percentile values).
6. Review of results by the experts and revision of initial answers by experts. This iterative reexamination of issues would sometimes increase the accuracy of results. Respondents who provide answers outside the interquartile range need to provide written justifications or arguments on the second cycle of completing the questionnaires.
7. Revision of results and review for another cycle. The process should be repeated until a complete consensus is achieved. Typically, the Delphi method requires about two cycles or iterations.
8. A summary of the results is prepared with argument summary for out of interquartile range values.

The responses on the final iteration usually show less spread in comparison to spreads in earlier iterations. The median values are commonly taken as the best estimates for the issues or questions.

The Delphi method offers the needed, adequate basis for expert-opinion elicitation; however, there is need to develop guidelines on its use to ensure consistency and result reliability. The remainder of this chapter and Chapter 4 provide the needed development and adaptation of this method for expert-opinion elicitation.

Example 3.1   Helmer (1968) Delphi questionnaire

This example provides a Delphi questionnaire as was originally developed and used by Helmer (1968). Table 3.1 shows the first of four parts of the questionnaire on technological innovations and use in the U.S. These questions were also used in the 1963 long-range forecasting study by RAND, and in 1966 using 23 RAND employees as participants. The differences among the results from the three studies ranged from 0 to 21 years, with an average of six years showing adequate consistency.

Example 3.2   Fallacy of civil defense strategic planning of the 1960s

Herman Kahn led several RAND studies that were funded by the U.S. Air Force on the effects of thermonuclear war and civil defense (Cooke, 1991). He later founded the Hudson Institute in New York. He articulated the strategic posture of *finite deterrence* and its upgrade to *credible first strike capability* for thermonuclear war (Kahn, 1960). Finite deterrence requires maintaining an ability to inflict unacceptable damage on an enemy after absorbing a surprise nuclear attack. This strategy can be augmented by counterforce measures to limit enemy-attack effects, for example, by building fallout shelters. By having enough counterforce measures with the ability to deliver and knock out enemy missiles before they are launched, a credible first strike capability is achieved. Kahn's argument includes the initiation of a nuclear war in the case of a desperate crisis or provocation that would be morally acceptable. A *desperate crisis* is defined as "a circumstance in which,

Table 3.1 Delphi Questionnaire

Questionnaire #1

- This is the first in a series of four questionnaires intended to demonstrate the use of the Delphi technique in obtaining reasoned opinions from a group of respondents.

- Each of the following six questions is concerned with developments in the United States within the next few decades.

- In addition to giving your answer to each question, you are also being asked to rank the questions from 1 to 7. Here "1" means that in comparing your own ability to answer this question with what you expect the ability of the other participants to be, you feel that you have the relatively best chance of coming closer to the truth than most of the others, while a "7" means that you regard that chance as relatively least.

Rank	Question	Answer*
☐	1. In your opinion, in what year will the median family income (in 1967 dollars) reach twice its present amount?	☐
☐	2. In what year will the percentage of electric automobiles among all automobiles in use reach 50%?	☐
☐	3. In what year will the percentage of households that are equipped with computer consoles tied to a central computer and data bank reach 50%?	☐
☐	4. By what year will the per-capita amount of personal cash transactions (in 1967 dollars) be reduced to one-tenth of what it is now?	☐
☐	5. In what year will the power generation by thermonuclear fusion become commercially competitive with hydroelectric power?	☐
☐	6. By what year will it be possible by commercial carriers to get from New York to San Francisco in half the time that is now required to make that trip?	☐
☐	7. In what year will a man for the first time travel to the moon, stay at least one month, and return to earth?	☐

* "Never" is also an acceptable answer.

- Please also answer the following question, and give your name (this is for identification purposes during the exercise only; no opinions will be attributed to a particular person).

  Check one: ☐ I would like to participate in the three remaining questionnaires.
  ☐ I am willing but not anxious to participate in the three remaining questionnaires.
  ☐ I would prefer not to participate in the three remaining questionnaires.

Name (block letters please): _____

Helmer, 1968

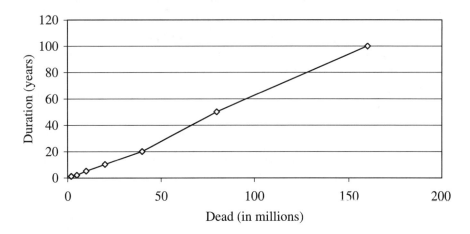

Figure 3.1 Estimated duration for thermonuclear postwar economic recuperation.

destabilizing as it would be, we would feel we would need an ability to rescue ourselves from a more dire eventuality by increasing our bargaining power or by actual use of the credible first strike capability" (Kahn, 1960).

The argument of RAND for credible first strike capability is based on expert opinion of the acceptable nature of retaliatory blow by an enemy, as demonstrated in Figure 3.1 in the form of an estimated duration in years for thermonuclear postwar economic recuperation. Kahn goes further to state ". . . Our calculations indicate that even without special stockpiling, dispersal, or protection, the restoration of our prewar gross national product should take place in a relatively short time — if we can hold the damage to the equivalent of something like 53 metropolitan areas destroyed." The results were based on the assumptions of "(1) favorable political environment (i.e., not losing the war), (2) immediate survival and patch-up, (3) maintenance of economic momentum, (4) specific bottlenecks alleviated, (5) *bourgeois* virtues survive, (6) workable postwar standards adopted, and (7) neglected effects unimportant." These are uncertain assumptions that were arguably justified by Kahn (1960) and were set at levels that were described as more likely to be pessimistic than optimistic.

The analysis by RAND did not adequately deal with uncertainty and ignorance. It weighed heavily cognitive knowledge and expert opinion creating overconfidence in the results. Newman (1961) provided a review in *Scientific America* of Kahn's book in which he conjectured that the entire book was a staff joke in poor taste (Newman, 1961; Freeman, 1969). The RAND study failed in properly assessing ignorance that places limits on human knowledge as shown in Figure 1.9. Since the publication of *Thermonuclear War* (Kahn, 1960), the phenomenon of electromagnetic pulse and potential climatological changes as a result of thermonuclear war were identified. These problems were not considered by RAND. The latter problem can result from the injection of millions of tons of dust and smoke in the upper atmosphere resulting in subfreezing land temperatures for months, and perhaps destroying human food resources such as crops. The effect of 100 to 10,000 total megatons of nuclear exchange could conceivably reduce the "population size of homosapians to prehistoric levels or below, and the extinction of human species itself cannot be excluded" (*Science*, 1983). Another failure of the RAND study is in logic used to conduct reasoning under uncertainty. For example, Kahn arguably concludes that after a small nuclear destruction scenario of 53 metropolitan areas, we will *probably* restore our gross national product (GNP) quickly. He argues that it is *likely* that we can handle radiation, it is *likely* that we can handle death, it is *likely* that we can handle destruction, therefore it is *likely* that we can handle jointly radiation, death and destruction. As a result, he concludes that we will *probably* restore our GNP quickly. A fallacy of this logic in probabilistic reasoning is that high probabilistic likeliness of three propositions does not necessarily lead to a high probabilistic likeliness of their joint proposition. Uncertainty does not propagate in this simple manner as was used by Kahn. A proper treatment of uncertainty through assessment, modeling, propagation and integration is essential in conjecture.

Example 3.3 NASA's Challenger space shuttle risk study

NASA sponsored a study to assess the risks associated with the space shuttle (Colglazier and Weatherwax, 1986; Cooke, 1991). In this study, an estimate of solid rocket booster failure probability per launch, based on subjective probabilities and operating experience, was estimated to be about 1 in 35. The probability was based on Bayesian analysis utilizing prior experience of 32 confirmed failures from 1902 launches of various solid rocket motors. This estimate was disregarded by NASA, and a number of 1 in 100,000 was dictated based on subjective judgments by managers and administrators (Colglazier and Weatherwax, 1986; Cooke, 1991). The dictated number was not in agreement with published data (Bell and Esch, 1989). The catastrophic Challenger explosion occurred on the twenty-fifth launch of a space shuttle on January 28, 1986.

Historically, NASA was distrustful of absolute reliability numbers for various reasons. It was publicized that the reliability numbers tend to be optimistic or taken as facts which they are not (Wiggins, 1985). In reality, failure probabilities can be threatening to the survival of NASA's mission programs. For example, a General Electric qualitative probabilistic study on the probability of successfully landing a man on the moon was 95%. NASA felt that such numbers could do an irreparable harm, and efforts of this type should be disbanded (Bell and Esch, 1989).

At the present, NASA is aggressively pursuing safety studies using probabilistic risk analysis of its various space missions. This change in NASA's practices can be attributed to the extensive investigations following the 1986 shuttle disaster.

NASA has used risk assessment matrices to avoid the problem of managers treating the values of probability and risk as absolute judgments

(Wiggins, 1985). The Department of Defense offers the use of risk assessment matrices as a tool to prioritize risk (Defense Acquisition University, 1998). Qualitatively, the likelihood of occurrence and consequences of an adverse scenario may be described as shown in Tables 3.2 and 3.3, respectively. Levels of occurrence may be based on expert-opinion elicitation or actual probability data. The consequences described in Table 3.3 can be determined using expert-opinion elicitation. Tables 3.2 and 3.3 can be combined to from the risk matrix. Risk assessment is based on the pairing of the likelihood of occurrence and consequences. Table 3.4 shows this pairing and is called a risk assessment matrix.

Table 3.2 Likelihood of Occurrence

Level	Description	Detailed Description
A	Frequent	Likely to occur frequently
B	Probable	Will occur several times in life of a system
C	Occasional	Likely to occur at sometime in life of a system
D	Remote	Unlikely but possible to occur in life of a system
E	Improbable	So unlikely that it can be assumed its occurrence may not be experienced

Wiggins, 1985

Table 3.3 Consequences

Level	Description	Mishap Definition
I	Catastrophic	Death or system loss
II	Critical	Severe injury, severe occupational illness, or major system damage
III	Marginal	Minor injury, minor occupational illness, or minor system damage
IV	Negligible	Less than minor injury, occupational illness, or system damage

Wiggins, 1985

Table 3.4 Risk Assessment Matrix Using a Risk Index

Likelihood level	Consequence level			
	I Catastrophic	II Critical	III Marginal	IV Negligible
A: Frequent	1	3	7	13
B: Probable	2	5	9	16
C: Occasional	4	6	11	18
D: Remote	8	10	14	19
E: Improbable	12	15	17	20

Criteria based on Risk Index:	1-5 = Unacceptable;
	6-9 = Undesirable (project management decision required);
	10-17 = Acceptable with review by project management; and
	18-20 = Acceptable without review.

Wiggins, 1985

3.3.2 Scenario analysis

The development of scenario analysis can be attributed to Kahn and Wiener (1967). A scenario is defined as a hypothetical sequence of events that are constructed to focus attention on causal processes and decision points or nodes. Scenario analysis attempts to answer two questions: (1) how might some hypothetical situation come about, step by step, and (2) what alternatives or choices exist for each actor or party to the situation, at each step, for preventing, diverting, or facilitating the process. The first question is addressed in a similar manner to what is called event tree analysis as described by Ayyub and McCuen (1997). The second question is commonly handled today using a decision tree as described by Ayyub and McCuen (1997). Kahn and Wiener (1967) used scenario analysis to predict technological innovations for the year 2000. An examination of their top likely 25 technological innovations would reveal a success rate of about 40%. The predictions are based on 50% occurrence likelihood.

The scenario analysis by Kahn and Wiener (1967) did not use scenario probabilities and relied on identifying what is termed the *surprise-free scenario* that is used as a basis for defining *alternative futures* or *canonical variations*. The alternative futures or canonical variations are generated by varying key parameters of the surprise-free scenario. Probabilities, that are absent from such an analysis, are arguably justified by Kahn and Wiener (1967) due to long-term projections making all scenarios of small likelihood. The surprise-free scenario is considered important due to its ability in defining the long-term trend rather than its likelihood. Therefore, it is important to bear in mind this limitation of scenario analysis, its inability to deliver likelihood predictions to us, but only long-term trends. At the present, this limitation can be alleviated by using event and decision tree analyses.

3.4 Scientific heuristics

The contemporary philosopher of science Hans Reichenbach (1951) made a distinction between discovery and justification in science. Discovery in science can be characterized as nonhomogenous, subjective, and nonrational. It can be based on hunches, predictions, biases, and imaginations. It is the product of creativity that extends in the domain of the unknown knowledge to humankind as shown in Figure 1.7. For example, everyone has seen the moon movement across the night sky and seen apples and other objects falling to earth; however, it took a Newton to realize the same physical laws underlie both phenomena. Newton's ideas were subjected to testing and validation using the scientific processes of justification. There is surely a difference between discovering ideas or phenomena and scientifically justifying them. The process of discovery and justification in science can be viewed as a rational consensus process that is based on empirical control (testing) and repeatability; i.e., the outcome of ideas should pass empirical testing by anyone and should be repeatable by anyone. *Heuristics* is a process of discovery that is not necessarily structured.

Discovery is a form of scientific heuristics that does not entail much structure and relies heavily on rules of thumb, subjectivity, and creativity. In order to be successful in its pursuit, it cannot approach issues at hand in an orderly fashion but requires a level of coherent disorder that must not reach a level of disarray. Subjectivity and disorder can lead to errors, especially biases, that are not intentional, although intentional or motivational biases can be present and should be targeted for elimination. Psychometric researchers such as Kahneman et al., (1982) and Thys (1987) have studied this area extensively on the fundamental level and to understand its relation to expert opinions, respectively.

Heuristics are the product of four factors that could lead to bias: (1) availability, (2) anchoring, (3) representativeness, and (4) control, as shown in Figure 3.2. For a given issue, availability is related to the ease with which individuals (including experts) can recall similar events or situations to this issue. Therefore, probabilities of well-publicized events tend to be overestimated, whereas probabilities of unglamorous events are underestimated.

Anchoring is the next factor in heuristics in which subjects, i.e., individuals or experts, tend to start with an initial estimate and correct it to the issue at hand. However, the correction might not be sufficient. For example, high school kids guessed the order of magnitude differences in estimating the product of the following two number sequences within a short period of time:

$$(8 \times 7 \times 6 \times 5 \times 4 \times 3 \times 2 \times 1) \text{ and } (1 \times 2 \times 3 \times 4 \times 5 \times 6 \times 7 \times 8)$$

The differences can be attributed to performing the first few multiplications, establishing anchors, and estimating the final answers through extrapolation (Kahneman et al., 1982).

Representativeness can affect conditional probability assessments. For example, individuals tend to evaluate intuitively the conditional probability $P(A|B)$ by assessing the similarity between A and B. The problem with this assessment is that similarity is symmetric whereas conditional probabilities are not; i.e., the resemblance of A to B is the same as the resemblance of B to A, whereas $P(A|B)$ does not equal $P(B|A)$.

The *control* factor refers to the perception of subjects in that they can control or had control over outcomes related to an issue at hand. For example, Langer (1975) demonstrated that lottery ticket buyers demanded higher median prices for reselling their tickets to a ticket seeker if they had selected the ticket numbers than others who were given tickets with randomly selected numbers. The false sense of control has contributed to a higher belief in the value of their tickets in comparison to other tickets.

Other sources of bias or error include base-rate fallacy and overconfidence. The base-rate fallacy arises as a result of using misguided or misinformed subjects. A subject might rely on recent or popular information and unintentionally ignore the historic rate for an event of interest. The recent or popular information might make the subject biased towards a substantially

Chapter three: Experts, opinions, and elicitation methods 107

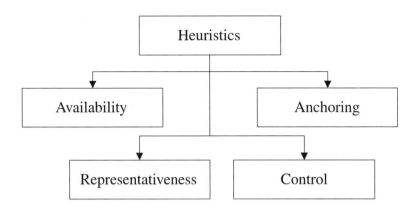

Figure 3.2 Factors affecting heuristics.

different rate than the historic value. For example, a subject might assign a relatively large occurrence probability for a failure event of a component as a result of recent or highly popularized failure information, in spite of historically low failure probabilities for such components. The base rate, which is low in this case, should be combined with the new information (i.e., recent or highly popularized failure information) using Bayes' theorem, resulting in a relatively small change in the base rate.

Overconfidence results in error and biases usually as a result of poor calibration (Cooke, 1991). Overconfidence is especially common in assessing confidence intervals on an estimated value. Subjects tend to provide narrower confidence intervals compared to real intervals. Studies have shown that the discrepancy between correct answers and overconfident answers increases as a respondent is more knowledgeable (Piattelli-Palmarini, 1994). Overconfidence is greatest in areas of expertise of a respondent. Calibration can help in controlling overconfidence by providing needed training. Overconfidence also appears in assessing small (or large) probabilities, less than 0.01 or in some cases less than 0.1 (larger than 0.99 or in some cases larger than 0.9).

Subject calibration can help in reducing the effects of base-rate fallacy and overconfidence. A well-calibrated subject can be defined as an individual who would consistently produce an estimate that is in agreement with the corresponding true value. Subjects can be calibrated by providing them with feedback on their assessments in training-like sessions. Expert calibration was successfully performed for weather forecasting, as was reported by Murphy and Daan (1984). The calibration process involves training subjects of probability concepts, error sources, biases, expectation, issue familiarization, aggregation methods, reporting, and use of results (Alpert and Raiffa, 1982; Murphy and Daan, 1984; Winkler and Murphy, 1968; Ferrell, 1994).

Subjectively assessed probabilities should be examined carefully for any signs of error or inadequacy. Historically, such signs include (1) data spread, (2) data dependence, (3) reproducibility, and (4) calibration. It is common to have spread in subjectively assessed probabilities, especially when dealing with low numbers (or large numbers for their complementary events). For example, the failure probability per hour of a high-quality steel pipe (10-m long) of a diameter at least 7.6 cm was subjectively assessed by 13 experts (Nuclear Regulatory Commission [NCR], 1975) as follows: 5E-6, 1E-6, 7E-8, 1E-8, 1E-8, 1E-8, 1E-8, 6E-9, 3E-9, 2E-9, 2E-10, 1E-10, and 1E-10, where 5E-6 mean 5×10^{-6}. The NRC used a value of 1E-10 with 90% confidence bounds of 3E-9 and 3E-12 in assessing the annual probability of core melt due to an earthquake. The following observations can be made based on these assessments:

1. The data have a spread of 5E-6 to 1E-10 which can be expressed as an upper-limit to lower-limit ratio of about 10,000.
2. The adopted value corresponds to the smallest value in this spread.
3. The 90% confidence bounds contain only 5 values out of the 13 gathered values.

Data spread is common in dealing with low numbers. Data spread can be reduced by asking subjects with extreme views to justify their values, and re-eliciting the data to establish a consensus on a tighter spread, although, it is common to just report and discuss data spread for the purpose of justifying and adopting a value with associated confidence bounds.

Data dependence can arise as a result of pessimistic or optimistic subjects, i.e., consistently biased subjects that provide low or high values, respectively, in comparison to corresponding true (long-term) values. Statistical tests can be performed on data dependence as a result of using biased subjects (Cooke, 1986).

Reproducibility of results can be examined by performing a benchmark study that would require several teams to perform independent analyses based on a common set of information about a system. Then, the results of analyses are compared for spread in the form of ratios of maximum to minimum values reported by the teams. Several benchmark studies of this type were performed (for example, Amendola, 1986; Brune et al., 1983). The Amendola (1986) study was structured in four stages using 10 teams from different European countries to assess the failure probability of a feedwater system. The four stages were:

1. The first stage involved blind, independent evaluation by the teams as an initial probabilistic analysis without information sharing among the teams. The spread ratio in the resulting probabilities is 25 (i.e., 8E-4 and 2E-2).
2. Fault tree analysis was independently performed by the teams resulting in a spread ratio of 36. Afterwards the teams met to produce one fault tree, but could not agree on a common one.
3. A common fault tree was assigned to the teams. The teams used their own data to produce the system failure probability. The spread ratio

Chapter three: Experts, opinions, and elicitation methods 109

> in the resulting probabilities is 9. This stage isolates the effect of data on the results.
> 4. A common fault tree and data were given to the teams. The teams used their analytical tools to produce the system failure probability. The spread ratio in the resulting probabilities is about 1. This stage isolates the effect of the analytical tools.

Having data with small spread and without dependence, that have been reproduced by several teams, does not mean that the data are correct; it only increases our confidence in them. The process of calibration is closely tied to the process of result validation which is difficult since opinion elicitation is commonly associated with rare events that cannot be validated. Training of subjects, however, can be based on other events or issues in order to have calibrated subjects.

Example 3.4   Information communication for national security intelligence

The intelligence community is in the business of information collection. It is quite common that gathered information is marred with subjectivity, uncertainty, and perhaps irrelevance, and it can be from nonreliable sources. The intelligence community is aware of and regularly deals with these problems. For example, the U.S. Defense Intelligence Agency (DIA) investigated uncertainty in intelligence information (Morris and D'Amore, 1980) and provided a summary of various conceptual and analytical models for this purpose. The primary interest of the study was to assess uncertainty in projecting future force levels of the former USSR. A secondary motive was to study the failure to predict the fall of the Shah of Iran in 1979 (Cooke, 1991).

The intelligence community widely used a reliability-accuracy rating system to communicate uncertainty, as shown in Table 3.5. However, Samet (1975) indicated that this system is not adequate, since correspondents tend to emphasize information accuracy, and does not necessarily convey uncertainty attributed to source reliability. The DIA used the Kent chart, as shown in Table 3.6, to provide a quantitative interpretation of natural language expressions of uncertainty. As reported by Morris and D'Amore (1980), however, the Kent chart has been replaced by a direct use of probabilities.

3.5   Rational consensus

The use of expert opinions in engineering and science needs to be performed as a part of a rational consensus process. A rational consensus process should meet the following requirements (Cooke, 1991):

- *Reproducibility.* The details of collection, gathering, and computation of results based on expert opinions need to be documents to a level

Table 3.5 Reliability and Accuracy Ratings in Intelligence Information

Source Reliability	Information Accuracy
A. Completely reliable	1. Confirmed
B. Usually reliable	2. Probably true
C. Fairly reliable	3. Possibly true
D. Not usually reliable	4. Doubtfully true
E. Unreliable	5. Improbable
F. Reliability cannot be judged	6. Accuracy cannot be judged

Morris and D'Amore, 1980

Table 3.6 A Kent Chart

Likelihood Order	Synonyms	Chances in 10	Percent
Near certainty	Virtually (almost) certain, we are convinced, highly probable, highly likely	9	99 to 90
Probable	Likely	8	60
	We believe	7	
	We estimate	6	
	Chances are good		
	It is probable that		
Even chance	Chances are slightly better than even	5	40
		4	
	Chances are about even		
	Chances are slightly less than even		
Improbable	Probably not	3	10
	Unlikely	2	
	We believe . . . not		
Near impossibility	Almost impossible	1	1
	Only a slight chance		
	Highly doubtful		

Note: Words such as "perhaps," "may," and "might" will be used to describe situations in the lower ranges of likelihood. The word "possible," when used without further modification, will generally be used only when a judgment is important but cannot be given an order of likelihood with any degree of precision.

Morris and D'Amore, 1980

that make them reproducibility by other expert peers. This requirement is in agreement with acceptable scientific research.
- *Accountability.* Experts, their opinion, and sources should be identified for reference by others as expert anonymity might degrade outcomes of consensus building and expert-opinion elicitation.
- *Empirical control.* Expert opinion should be susceptible to empirical control if possible at a minimum for selected practical cases. Empirical control can be performed by comparing results of expert-opinion elicitation with observations for selected control issues.

This empirical control might not be possible in some situations, but it is in agreement with acceptable scientific research.
- *Neutrality.* The method of eliciting, evaluating, and combining expert opinions should encourage experts to state their true opinions. For example, the use of the median to aggregate expert opinions might violate this requirement if the median is perceived to reward centrally compliant experts. Methods of using weighted averages of opinions based on self weights or weights by experts of each other have the same fallacy.
- *Fairness.* The experts should be equally treated during the elicitation and for the purposes of processing the observations.

3.6 Elicitation methods

This section provides a summary of various methods that can be used for elicitation of expert opinions. In order to increase the chances of success in using elicitation and scoring methods, Cooke (1991) provided suggested practices and guidelines. They were revised for the purposes herein, and are summarized as follows:

1. The issues or questions should be clearly stated without any ambiguity. Sometimes there might be a need for testing the issues or questions to ensure their adequate interpretation by others.
2. The questions or issues should be stated using appropriate format with listed answers, perhaps graphically expressed, in order to facilitate and expedite the elicitation and scoring processes.
3. It is advisable to test the processes by performing a dry run.
4. The analysts must be present during the elicitation and scoring processes.
5. Training and calibration of experts must be performed. Examples should be presented with explanations of elicitation and scoring processes, and aggregation and reduction of results. The analysts should avoid coaching the experts or leading them to certain views and answers.
6. The elicitation sessions should not be too long. In order to handle many issues, several sessions with appropriate breaks might be needed.

3.6.1 Indirect elicitation

The indirect elicitation method is popular among theoreticians and was independently introduced by Ramsey (1931) and De Finetti (1937). The indirect method is based on betting rates by experts in order to reach a point of indifference among presented options related to an issue. The primary disadvantage of this method is the utility value of money is not necessarily linear with the options presented to an expert, and the utility value of money is independent of the answer to an issue, such as failure rate.

Other indirect techniques were devised by researchers in order to elicit probabilities from probability-illiterate experts. For example, analysts have used time to first failure estimation or age at replacement for a piece of equipment as an indirect estimation of failure probability.

Example 3.5 Betting rates for elicitation purposes

Betting rates can be used to subjectively and indirectly assess the occurrence probability of an event A, called $p(A)$. According to this method, an expert E is hypothetically assigned a lottery ticket of the following form:

$$\text{Expert } E \text{ receives } \$100 \text{ if } A \text{ occurs.} \tag{3.1}$$

The interest hereafter becomes the value that the expert attaches to this lottery ticket. For an assumed amount of money $\$x$, that is less than $\$100$, the expert is asked to trade the ticket for the $\$x$ amount. The amount $\$x$ is increased incrementally until a point of indifference is reached; i.e., the lottery ticket has the same value as the offered $\$x$ amount. The $\$x$ position is called *certainty equivalent* to the lottery ticket.

Assuming the expert to be a rational and unbiased agent, the $\$x$ position which is certainty equivalent to the lottery ticket, provides an assessment of an expectation. The expected utility of the lottery ticket can be expressed as

$$\text{Expected utility of the lottery ticket} = \$100k(p(A)) \tag{3.2}$$

where $p(A)$ = the occurrence probability of A, and k = a constant that represent the utility for money as judged by the expert. The utility of money can be a nonlinear function of the associated amount. At the certainty equivalent position, $\$x$ has a utility of $k\$x$, which is equivalent to the expected utility of the lottery ticket as shown in Equation 3.2. Therefore, the following condition can be set:

$$(\$100)k(p(A)) = k(\$x). \tag{3.3}$$

Solving for $p(A)$ produces

$$p(A) = \frac{\$x}{\$100}. \tag{3.4}$$

The utility of money in the above example was assumed to be linear, whereas empirical evidence suggests that it is highly nonlinear. Galanter (1962) constructed Table 3.7 by asking subjects the following question:

> "Suppose we give you x dollars; reflect on how happy you are. How much should we have given you in order to have made you twice as happy?"

Chapter three: Experts, opinions, and elicitation methods

Table 3.7 Money Required to Double Happiness

Given x	Twice as Happy	
	Mean	Median
$10	$53.	$45
$20	$538	$350
$1000	$10,220	$5000

Galanter, 1962

The following utility function U was developed based on these data:

$$U(x) = 3.71x^{0.43} \tag{3.5}$$

It is evident that the willingness of people to run a risk does not grow linearly with an increased amount x. Similar tests were performed for losses of money and their relationship to unhappiness, but were inconclusive as subjects found the questions "too difficult." Therefore, betting rates might not be suitable for failure probability assessment especially since such probabilities are commonly very small.

3.6.2 Direct method

This method elicits a direct estimate of the degree of belief of an expert on some issue. Despite its simple nature, this method might produce the worst results, especially from experts who are not familiar with the notion of probability. Methods that fall in this category are the Delphi method and the nominal group technique. The Delphi technique, as described in detail in Section 3.3.1, allows for no interaction among the elicited expert before rendering opinions. Variations to this method were used by engineers and scientists by allowing varying levels of interactions that range from limited interaction to complete consensus building. The nominal group technique allows for a structured discussion after the experts have provided initial opinions. The final judgment is made individually on a second cycle of opinion elicitation and aggregated mathematically similar to the Delphi method (Gustafson et al., 1973; Morgan and Henrion, 1992). Lindley (1970) suggested a method that is based on comparing an issue to other familiar issues with known answers. This comparative examination has been proven to be easier for experts than directly providing absolute final answers. For example, selected experts that are familiar with an event A and its occurrence probability $p(A)$ are accustomed to assessing subjectively the occurrence probability of event B. We are interested in assessing the occurrence probability of event B that is not of the same probability familiarity to the experts as $p(A)$. Experts are asked to assess the relative occurrence of B to A, say 10 times as frequent. Therefore, $p(B) = 10p(A)$ as long as the result is less than one.

3.6.3 Parametric estimation

Parametric estimation is used to assess the confidence intervals on a parameter of interest such as the mean value. The estimation process can be in the form of a two-step procedure as follows (Preyssl and Cooke, 1989):

1. Obtain a median estimate of a probability (m), and
2. Obtain the probability (r) that the true value will exceed 10 times the median value (m).

The m and r values can be used to compute the 5% and 95% confidence bounds as $m/k_{0.95}$ and $m(k_{0.95})$, respectively, where

$$k_{0.95} \approx \frac{\exp(-0.658)}{z_{1-r}} \tag{3.6}$$

in which z_{1-r} is the $(1-r)^{\text{th}}$ quantile value of the standard normal probability distribution. Some experts were found to like and favor two-step methods for dealing with uncertainty.

3.7 Standards for educational and psychological testing

Credible behavioral testing and research adhere to the Standards for Educational and Psychological Testing (SEPT) published by the American Psychological Association (1985). The objective of this section is to summarize these standards, to determine how they relate to expert-opinion elicitation, and to identify any pitfalls in expert-opinion elicitation based on examining these standards.

Sacman (1975) from the RAND Corporation provided a highly critical critique of the Delphi methods based on its lack of compliance with the SEPT, among other scientific and research practices. This critique is valuable and is summarized herein since its applicability in some concerns goes beyond the Delphi methods to other expert-opinion elicitation methods. Sacman (1975) found that conventional Delphi applications

- often involve crude questionnaire designs,
- do not adhere to proper statistical practices of sampling and data reduction,
- do not provide reliability measures,
- do not define scope, populations, and limitations,
- provide crisply stated answers to ambiguous questions,
- involve confusing aggregation methods of expert opinions with systematic predictions,
- inhibit individuality, encourage conformity, and penalize dissidents,
- reinforce and institutionalize early closure on issues,

Chapter three: Experts, opinions, and elicitation methods 115

- can give an exaggerated illusion of precision, and
- lack professional accountability.

Although his views are sometimes overstated, they are still useful in highlighting pitfalls and disadvantages of the Delphi method. The value of the Delphi method comes from its initial intended use as a heuristic tool, not a scientific tool, for exploring vague and unknown future issues that are otherwise inaccessible. It is not a substitute for scientific research.

According to the SEPT, a test involves several parties, as follows: test developer, test user, test taker, test sponsor, test administrator, and test reviewer. In expert-opinion elicitation studies, similar parties can be identified. The SEPT provide criteria for the evaluation of tests, testing practices, and the effects of test use. The SEPT provide a frame of reference to supplement professional judgment for assessing the appropriateness of a test application. The standard clauses of the SEPT are classified and identified as *primary standards* that should be met by all tests and *secondary standards* that are desirable as goals but are likely to be beyond reasonable expectation in many situations. The SEPT consist of four sections as follows:

Part I.   Technical Standards for Test Construction and Evaluation
Part II. Professional Standards for Test Use
Part III. Standards for Particular Applications
Part IV. Standards for Administrative Procedures

These SEPT parts are described in subsequent sections as they relate to expert-opinion elicitation.

Part I. Technical Standards for Test Construction and Evaluation
Part I of the SEPT provides standards for test construction and evaluation that contain standards for validity, reliability, test development, scaling, norming, comparability, equating, and publication.

The validity consideration of the SEPT covers three aspects: (1) construct-related evidence, (2) content-related evidence, and (3) criterion-related evidence. Construct-related evidence primarily focuses on the test score appropriateness in measuring the psychological characteristic of interest. In these guidelines, expert-opinion elicitation deals with occurrence likelihood and consequences. The corresponding test scores can be selected as probabilities and consequence units such as dollars. The use of these scores does meet the validity standards of SEPT in terms of a construct-related evidence. The content-related evidence requires that the selected sample is representative of some defined universe. In the context of expert-opinion elicitation, experts should be carefully selected in order to meet the content-related evidence. The criterion-related evidence needs to demonstrate that the test scores are related to a criterion of interest in the real world. In the context of expert-opinion elicitation, the estimated occurrence probabilities and consequences need to be related to corresponding real, but unknown, values. This criterion-

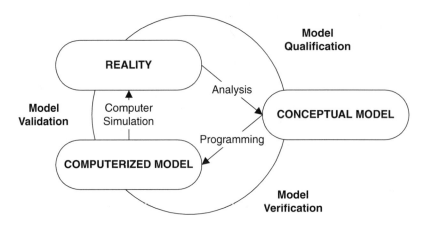

Figure 3.3 Verification and validation of computational fluid dynamics simulations (AIAA, 1998).

related evidence for validity is in agreement with the validation concept in the AIAA Guide for Verification and Validation of Computational Fluid Dynamics Simulations, as shown in Figure 3.3. The last consideration in validity is *validity generalization* that was reported in the form of the following two uses: (1) to draw scientific conclusions, and (2) to transport the result validity from one case to another. In the context of expert-opinion elicitation, validity generalization based on these two uses might be difficult to justify. Selected primary validity standards, most related to expert-opinion elicitation, are shown in Table 3.8. They were taken from the 1997 draft revision of the SEPT posted on the World Wide Web site of the American Psychological Association.

The reliability consideration of the SEPT deals with measurement errors due to two primary sources: (1) variations from one subject to another that are subjected to the same conditions and provided with the same background information, and (2) variations from one occasion to another by the same subject. The tools that are needed to estimate the reliability of the scores and test measurement errors are dependent on the error type. Statistical methods can be used for this purpose. In the context of expert-opinion elicitation, this reliability consideration requires aggregation procedures of expert opinions to include measures of central tendency, biases, dispersion, correlation, variances, standard error of estimates, spread of scores, sample sizes, and population definition.

Part I of the SEPT requires that tests and testing programs should be developed on a sound scientific basis. The standards puts the responsibility on the test developers and publishers to compile evidence bearing on a test, decide which information is needed prior to test publication or distribution and which information can be provided later, and conduct the necessary research.

Chapter three: Experts, opinions, and elicitation methods 117

Table 3.8 Selected Validity Standards from the Standards for Educational and Psychological Testing

1997 Draft SEPT Standard	Standard Citation	Relationship to Expert-Opinion Elicitation
1.1	A rationale should be presented for each recommended interpretation and use of test scores, together with a comprehensive summary of the evidence and theory bearing on the intended use or interpretation.	Definition of issues for expert-opinion elicitation.
1.2	The test developer should set forth clearly how test scores are intended to be interpreted and used. The population(s) for which a test is appropriate should be clearly delimited, and the construct that the test is intended to assess should be clearly described.	Definition of issues for expert-opinion elicitation.
1.3	If validity for some common or likely interpretation has not been investigated or is inconsistent with available evidence, that fact should be made clear, and potential users should be cautioned about making unsupported interpretations.	Definition of issues for expert-opinion elicitation.
1.4	If a test is used in a way other than those recommended, it is incumbent on the user to justify the new use, collecting new evidence if necessary.	Definition of issues for expert-opinion elicitation.
1.5	The composition of any sample of examinees from which validity evidence is obtained should be described in as much detail as is practicable, including major relevant sociodemographic and developmental characteristics.	Selection of and training of experts.
1.7	When a validation rests in part on the opinions or decisions of expert judges, observers, or raters, procedures for selecting such experts and for eliciting judgments or ratings should be fully described. The qualifications and experience of the judges should be presented.	Selection of and training of experts, and definition of aggregation procedures of expert opinions.

APA, 1997

The scaling, norming, comparability, and equating considerations in the SEPT deal with aggregation and reduction of scores. The documentation of expert-opinion elicitation should provide experts and users with clear explanations of the meaning and intended interpretation of derived score scales, as well as their limitations. Measurement scales and aggregation methods with their limitations, that are used for reporting scores, should be clearly described in expert-opinion elicitation documents. The documents should also include clearly defined populations that are covered by the expert-opinion elicitation process. For studies that involve score equivalence or comparison and equating of findings, detailed technical information should be provided on equating methods or other linkages and on the accuracy of equating methods.

Administrators of a test should publish sufficient information on the tests in order for qualified users and reviewers to reproduce the results and/or assess the appropriateness and technical adequacy of the test.

Part II. Professional Standards for Test Use
Part II of the SEPT provides standards for test use. Users of the results of a test should be aware of methods used in planning, conducting, and reporting the test in order to appreciate the limitations and scope of use of the test. Documented information on validity and reliability of test results as provided in Part I of the SEPT should be examined by the users for this purpose.

This part also deals with clinical testing, educational and psychological testing at schools, test use in counseling, employment testing, professional and occupational licensure and certification, and program evaluation. These standards have minimal relevance to expert-opinion elicitation.

Part III. Standards for Particular Applications
Part III of the SEPT provides standards for testing linguistic minorities and people with handicapping conditions. These standards have minimal relevance to expert-opinion elicitation.

Part IV. Standards for Administrative Procedures
Part IV of the SEPT provides standards for test administration, scoring, reporting, and rights of test takers. This part requires that tests should be conducted under standardized and controlled conditions similar to conducting experimental testing. Standardized and controlled conditions enhance the interpretation of test results by increasing the interpretation quality and effectiveness. This part also deals with access to test scores, i.e., test security and cancellation of test scores because of test irregularities.

3.8 Methods of social research

Social research concerns itself with gathering data on specific questions, issues, or problems of various aspects of society and thus helps humans to understand society. Social study has evolved to social science, especially in

Chapter three: Experts, opinions, and elicitation methods 119

the field of sociology where there are three primary schools of thought (Bailey, 1994):

1. humans have free will, and thus no one can predict their actions and generalize about them (the Wilhelm Dilthey school of the 19th century),
2. social phenomena are orderly and can be generalized, and they adhere to underlying social laws that need to be discovered through research similar to physical laws (the Emile Durkheim methods of *positivisim*), and
3. social phenomena are the product of free-will human volitional actions that are not random and can be predicted by understanding the human rationale behind them (an intermediate school of thought of Max Weber).

The stages of social research can be expressed in a circle of five stages as shown in Figure 3.4 to allow for feedback in redefining a hypothesis in the first stage.

The construction and use of questionnaires is common and well developed in social research. Experiences from this field might be helpful to developers, facilitators and administrators of expert-opinion elicitation processes. The construction of a questionnaire should start by defining its relevance at the following three levels:

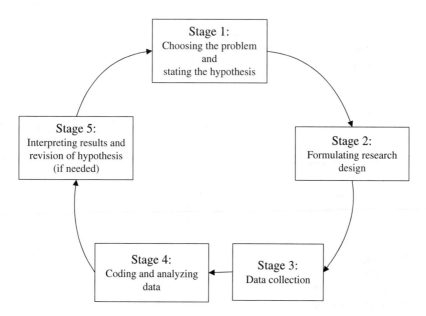

Figure 3.4 Stages of social research.

1. **Relevance of the study to the subjects:** It is important to communicate the goal of the study to the subjects, and establish its relevance to them. Establishing this relevance would make them stake holders and thereby increase their attention and sincerity levels.
2. **Relevance of the questions to the study:** Each question or issue in the questionnaire needs to support the goal of the study. This question-to-study relevance is essential to enhancing the reliability of collected data from subjects.
3. **Relevance of the questions to subjects:** Each question or issue in the questionnaire needs to be relevant to each subject, especially when dealing with subjects of diverse views and backgrounds.

The following are guidelines on constructing questions and stating issues:

- Each item on the questionnaire should include only one question. It is a poor practice to include two questions in one.
- Question or issue statements should not be ambiguous. Also, the use of ambiguous words should be avoided. In expert-opinion elicitation of failure probabilities, the word "failure" might be vague or ambiguous to some subjects. Special attention should be given to its definition within the context of each issue or question.
- The level of wording should be kept to a minimum. Long questions should be avoided. Also, the word choice might affect the connotation of an issue especially by different subjects. Words should be selected carefully to meet the goal of the study in a most reliable manner.
- The use of factual questions is preferred over abstract questions. Questions that refer to concrete and specific matters result in desirable concrete and specific answers.
- Questions should be carefully structured in order to reduce biases of subjects. Questions should be asked in a neutral format, sometimes more appropriately without lead statements.
- Sensitive topics might require stating questions with lead statements that would establish supposedly accepted social norms in order to encourage subjects to answer the questions truthfully.

Questions can be classified as *open-ended* and *closed-ended*. A closed-ended question has the following characteristics:

- limits the possible outcomes of response categories,
- can provide guidance to subjects thereby making it easier for the subjects,
- provides complete answers,
- allows for dealing with sensitive or taboo topics,
- allows for comparing the responses of subjects,
- produces answers that can be easily coded and analyzed,

- can be misleading,
- allows for guesswork by ignorant subjects,
- can lead to frustration due to subject perception of inappropriate answer choices,
- limits the possible answer choices,
- does not allow for detecting variations in question interpretation by subjects,
- results in artificially small variations in responses due to limiting the possible answers, and
- can be prone to clerical errors by subjects in unintentionally selecting wrong answer categories.

An open-ended question has the following characteristics:

- does not limit the possible outcomes of response categories,
- is suitable for questions without known answer categories,
- is suitable for dealing with questions with too many answer categories,
- is preferred for dealing with complex issues,
- allows for creativity and self expression,
- can lead to collecting worthless and irrelevant information,
- can lead to nonstandardized data that cannot be easily compared among subjects,
- can produce data that are difficult to code and analyze,
- requires superior writing skills,
- might not communicate properly the dimensions and complexity of the issue,
- can be demanding on the time of subjects, and
- can be perceived as difficult to answer and thereby discourages subjects from responding accurately or at all.

The format, scale, and units for the response categories should be selected to best achieve the goal of the study. The minimum number of questions and question order should be selected with the following guidelines:

- sensitive questions and open-ended questions should be left to the end of the questionnaire,
- the questionnaire should start with simple questions that are easy to answer,
- a logical order of questions should be developed such that questions at the start of the questionnaire feed needed information into questions at the end of the questionnaire,
- questions should follow other logical orders that are based on time-sequence or process,
- the order of the questions should not lead or set the response,

- reliability-check questions that are commonly used in pairs (stated positively and negatively) should be separated by other questions,
- questions should be mixed in terms of format and type in order to maintain the interest of subjects, and
- the order of the questions can establish a *funnel* by starting with general questions followed by more specific questions within several branches of questioning; this funnel technique might not be appropriate in some applications, and its suitability should be assessed on a case-by-case basis.

The final stage of developing a questionnaire is writing a cover letter or introductory statement, instructions to interviewers, subjects, or facilitators, precoding, and pretesting. The introductory statement should provide the goal of the study and establish relevance. The instructions should provide guidance on expectations, completion of questionnaire, and reporting. Precoding assigns numerical values to responses for the purpose of data analysis and reduction. Pretesting should be administered to a few subjects for the purpose of identifying and correcting flaws.

Some of the difficulties or pitfalls of using questionnaires, with suggested solutions or remedies, include the following (Bailey, 1994):

1. Subjects might feel that the questionnaire is not legitimate and has a hidden agenda. A cover letter or a proper introduction of the questionnaire is needed.
2. Subjects might feel that the results will be used against them. Unnecessary sensitive issues and duplicate issues should be removed. Sometimes assuring a subject's anonymity might provide the needed remedy.
3. Subjects might refuse to answer questions on the basis that they have done their share of questionnaires or are tired of "being a guinea pig." Training and education might be needed to create the proper attitude.
4. A "sophisticated" subject who participated in many studies may have thereby developed an attitude of questioning the structure of the questionnaire, test performance, and result use and might require a "sampling around" to find a replacement subject.
5. A subject might provide "normative" answers, i.e., answers that the subject thinks are being sought. Unnecessarily sensitive issues and duplicate issues should be removed. Sometimes assuring a subject's anonymity might provide the needed remedy.
6. Subjects might not want to reveal their ignorance and perhaps appear stupid. Emphasizing that there are no correct or wrong answers, and assuring a subject's anonymity, might provide the needed remedy.
7. A subject might think that the questionnaire is a waste of time. Training and education might be needed to create the proper attitude.
8. Subjects might feel that a question is too vague and cannot be answered. The question should be restated so that it is very clear.

3.9 Focus groups

The concept of focus groups was started in the 1930s when social scientists began questioning the appropriateness and accuracy of information that resulted from individual subject interviews using questionnaires with closed-ended response choices; they began investigating alternative ways for conducting interviews. Rice (1931) expressed his concern of having a questioner taking the lead, with subjects taking a more or less passive role; thereby resulting in not capturing information of the highest value as a result of the questioner's, intentionally or unintentionally, leading the subjects away. As a result, information gathered at interviews might embody preconceived ideas of the questioner as well as the attitude of the subjects interviewed. During World War II, social scientists started using nondirective interview techniques in groups as a start of focus groups. The first use of focus groups was by the War Department to assess morale of the U.S. military. It was noted that people tend to reveal sensitive information once they are in a safe, comfortable environment with people like themselves, as described in the classic book on focus groups by Merton, et al. (1956).

Focus groups are commonly used in assessing markets, product developments, developing price strategies for new products and their variations, and assessing some issues of interest. A focus group can be defined as a special type of group, in terms of purpose, size, composition, and procedures, that has a purpose of listening and gathering information based on how people feel or think about an issue (Krueger and Casey, 2000). The intent of focus groups is to promote self-disclosure among the subjects by revealing what people really think and feel.

Focus groups are popular in terms of use by commercial business, foundations, governmental agencies, and schools. The size of the group can be approximately six to twelve subjects, depending on the application and issues considered. For market research, the subjects should not know each other, preferably being strangers, but it is not an issue in other types of research. Each focus group should have a professional or qualified moderator. The focus groups should be held in special rooms with one-way mirrors and quality acoustics for market research, but they can be held in regular meeting rooms or a lounge for other applications. A list of questions should be prepared that generally can be structured as (1) opening questions, (2) introductory questions, (3) transition questions, (4) key questions, (5) ending questions. The questions should be structured and stated so that they are engaging to participants. The questions should be developed by a group of analysts through brainstorming, phrasing of questions, sequencing the questions, estimating time needed for questions, getting feedback from others, and testing the questions. The data can be captured through observation behind mirrors, audio and video recording, and/or field notes. The results can be analyzed in various manners depending on the issue or application. The analysis methods can range from rapid first impression by a moderator or an analyst, to transcripts followed by rigorous analysis using analytical and statistical methods, to abridged transcripts with charts and audiotapes. A report should be prepared documenting the process and results. The report should then be sent to the sponsor (in the case of market research), can be provided to public officials, can be shared with the participants, or can be shared with a community. The time needed to complete a focus group study is about two weeks in the case of market research to several months in other applications.

Krueger and Casey (2000) provide variations and tips to the above general model for convening and using focus groups for special cases, such as dealing with existing organizations, young people, ethnic or minority groups, international groups or organizations, periodically repeated groups, telephone and Internet focus groups, and media focus groups.

3.10 Exercise problems

Problem 3.1 What are the primary pitfalls of using the Delphi method? Provide examples.

Problem 3.2 What are the advantages of using the Delphi method? When would you use a Delphi type of a method? Provide examples.

Problem 3.3 What are the differences between the Delphi method and scenario analysis? Provide examples to demonstrate the differences. Can you provide an example in which both the Delphi method and scenario analysis should be used?

Problem 3.4 What are the factors that could bias scientific heuristics results? Provide examples.

Problem 3.5 What is the base-rate fallacy? How can it lead to an opinion in error?

Problem 3.6 Using the method of parametric estimation, compute the 5% and 95% confidence bounds on a concrete strength parameter based on a median (m) of 3000 ksi and the probability (r) of 0.01.

Problem 3.7 Using the method of parametric estimation, compute the 5% and 95% confidence bounds on a parameter based on a median (m) of 100 and the probability (r) of 0.02.

Problem 3.8 Construct a questionnaire that meets the Standards for Educational and Psychological Testing (SEPT) and social research to determine the meaning of unethical conduct in exam taking by students, and of student preferences, and probabilities of cheating, when taking in-class and take-home exams.

Problem 3.9 Construct a questionnaire that meets the SEPT and social research to determine the meaning of unethical conduct in engineering workplaces by engineers and probabilities of its occurrence.

Problem 3.10 What are the differences between a focus group and a Delphi session?

chapter four

Expressing and modeling expert opinions

Contents

4.1. Introduction	126
4.2. Set theory	127
4.2.1. Sets and events	127
4.2.2. Fundamentals of classical set theory	127
4.2.2.1. Classifications of sets	127
4.2.2.2. Subsets	128
4.2.2.3. Membership (or characteristic) function	128
4.2.2.4. Sample space and events	128
4.2.2.5. Venn-Euler diagrams	129
4.2.2.6. Basic operations on sets	130
4.2.3. Fundamentals of fuzzy sets and operations	131
4.2.3.1. Membership (or characteristic) function	132
4.2.3.2. Alpha-cut sets	134
4.2.3.3. Fuzzy Venn-Euler diagrams	135
4.2.3.4. Fuzzy numbers, intervals, and arithmetic	137
4.2.3.5. Operations on fuzzy sets	145
4.2.3.6. Fuzzy relations	152
4.2.3.7. Fuzzy functions	155
4.2.4. Fundamental of rough sets	156
4.2.4.1. Rough set definitions	156
4.2.4.2. Rough set operations	157
4.2.4.3. Rough membership functions	159
4.2.4.4. Rough functions	160
4.3. Monotone measures	162
4.3.1. Definition of monotone measures	163
4.3.2. Classifying monotone measures	164

        4.3.3. Evidence theory 165
            4.3.3.1. Belief measure 165
            4.3.3.2. Plausibility measure 166
            4.3.3.3. Basic assignment 166
        4.3.4. Probability theory 170
            4.3.4.1. Relationship between evidence theory
                     and probability theory 170
            4.3.4.2. Classical definitions of probability 171
            4.3.4.3. Linguistic probabilities 173
            4.3.4.4. Failure rates 173
            4.3.4.5. Central tendency measures 173
            4.3.4.6. Dispersion (or variability) 176
            4.3.4.7. Percentiles 177
            4.3.4.8. Statistical uncertainty 178
            4.3.4.9. Bayesian methods 181
            4.3.4.10. Interval probabilities 188
            4.3.4.11. Interval cumulative distribution functions 189
            4.3.4.12. Probability bounds 193
        4.3.5. Possibility theory 199
    4.4. Exercise problems 201

4.1 Introduction

We seek experts for their opinions on issues identified as of interest to us. Appropriate definitions of these issues and formats for anticipated judgments are critical for the success of an expert-opinion elicitation process. This chapter provides background information and analytical tools that can be used to express expert opinions and model these opinions. The expression of a view or judgment by an expert is an important link in the success of the process. The *expression* of an opinion can be defined as putting it into words or numbers, or representing the opinion in language, a picture, or a figure. The manner of expressing the opinion in a meaningful and eloquent manner is an important characteristic. The expression might be sensitive to the choice of a particular word, phrase, sentence, symbol, or picture. It can also include a show of feeling or character. It can be in the form of a symbol or set of symbols expressing some mathematical or analytical relationship, as a quantity or operation.

In this chapter, we present the fundamentals of classical set theory, fuzzy sets, and rough sets that can be used to express opinions. Basic operations for these sets are defined and demonstrated. Fuzzy relations and fuzzy arithmetic can be used to express and combine collected information. The fundamentals of probability theory, possibility theory, interval probabilities, and monotone measures will be summarized as they relate to the expression of expert opinions. Examples are used in this chapter to demonstrate the various methods and concepts.

4.2 Set theory

4.2.1 Sets and events

Sets constitute a fundamental concept in probabilistic analysis of engineering problems. Informally, a set can be defined as a collection of elements or components. Capital letters are usually used to denote sets, e.g., $A, B, X, Y, \ldots$, etc. Small letters are commonly used to denote their elements, e.g., $a, b, x, y, \ldots$, etc., respectively.

4.2.2 Fundamentals of classical set theory

The fundamental difference between classical (crisp or nonfuzzy) sets and fuzzy sets is that belonging and nonbelonging to sets are assumed to be without any form of uncertainty in the former and with uncertainty in the latter. This assumption that elements belong or do not belong to sets without any uncertainty constitutes a basis for the classical set theory. If this assumption is relaxed to allow some form of uncertainty in belonging or nonbelonging to sets, then the notion of fuzzy sets can be introduced where belonging and nonbelonging involve variable levels of uncertainty.

Example 4.1 Selected crisp sets

The following are example sets:

$$A = \{2, 4, 6, 8, 10\} \tag{4.1}$$

$$B = \{b: b > 0 \}; \text{ where ``:'' means ``such that''} \tag{4.2}$$

$$C = \{\text{Maryland, Virginia, Washington}\} \tag{4.3}$$

$$D = \{P, M, 2, 7, U, E\} \tag{4.4}$$

$$F = \{1, 2, 3, 4, 5, 6, 7, \ldots\}; \text{ the set of odd numbers} \tag{4.5}$$

In these example sets, each set consists of a collection of elements. In set A, 2 belongs to A, and 12 does not belong to A. Using mathematical notations, this can be expressed as $2 \in A$, and $12 \notin A$, respectively.

4.2.2.1 Classifications of sets

Sets can be classified as *finite* and *infinite* sets. For example, sets A, C, and D in Example 4.1 are finite sets, whereas sets B and F are infinite sets. The elements of a set can be either *discrete* or *continuous*. For example, the elements of sets A, C, D, and F are discrete, whereas the elements in set B are continuous. A set without any elements is named the null (or empty) set $\emptyset$.

A set that contains all possible elements is called the universe or universal space. Also, sets can be classified as *convex* and *nonconvex* sets. A convex set has the property that any straight line connecting any two elements of the set does not cross and go outside the set's boundaries.

4.2.2.2 Subsets

If every element in a set A is also a member in a set B, then A is called a subset of B. Mathematically expressed, $A \subset B$ if for every $x \in A$ implies $x \in B$. Every set is considered to be a subset of itself. The null set $\emptyset$ is considered to be a subset of every set.

Example 4.2 Example subsets

The following are example subsets:

$$A_1 = \{2, 4\} \text{ is a subset of } A = \{2, 4, 6, 8, 10\} \tag{4.6}$$

$$B_1 = \{b: 7 < b \leq 200\} \text{ is subset of } B = \{b: b > 0\} \tag{4.7}$$

$$F = \{1, 2, 3, 4, 5, 6, 7, \ldots\} \text{ is a subset of } F = \{1, 2, 3, 4, 5, 6, 7, \ldots\} \tag{4.8}$$

4.2.2.3 Membership (or characteristic) function

Let X be a universe, or a set of x values, and let A be a subset of X. Each element, x, is associated with a membership value to the subset A, $\mu_A(x)$. For an ordinary, nonfuzzy, or crisp set, the membership function is given by

$$\mu_A : X \to \{0,1\} \tag{4.9}$$

as a mapping from the universe X to the integer values $[0,1]$; where 0 means a value x does not belong to A, and 1 means a value x belongs to A. Also, the membership function can be expressed as

$$\mu_A(x) = \begin{cases} 1 & \forall x \in A \\ 0 & \forall x \notin A \end{cases} \tag{4.10}$$

where $\forall$ means for all. The meaning of this membership function is that there are only two possibilities for an element x, either being a member of A, i.e., $\mu_A(x) = 1$, or not being a member of A, i.e., $\mu_A(x) = 0$. In this case the set A has sharp boundaries.

4.2.2.4 Sample space and events

In probability theory, the set of all possible outcomes of a system constitute the sample space. A sample space consists of sample points that correspond to the possible outcomes. A subset of the sample space is called an event.

Chapter four: Expressing and modeling expert opinions   129

These definitions form the set-basis of probabilistic analysis. An event without sample points is an empty set and is called the impossible set Ø. A set that contains all the sample points is called the certain event S. The certain event is equal to the sample space.

Example 4.3   Example sample spaces and events

The following are example sample spaces:

$$A = \{a: a = \text{number of vehicles making a left turn at a specified traffic light within an hour}\} \quad (4.11)$$

$$B = \{b: b = \text{number of products produced by an assembly line within an hour}\} \quad (4.12)$$

$$C = \{\text{the strength of concrete delivered at a construction site}\} \quad (4.13)$$

Based on the sample space A, the following events can be defined:

$$A_1 = \{\text{number of cars making a left turn at the specified traffic light during one traffic light cycle}\} \quad (4.14)$$

$$A_2 = \{\text{number of vehicles making a left turn at the specified traffic light during two traffic light cycles}\} \quad (4.15)$$

4.2.2.5   *Venn-Euler diagrams*

Events and sets can be represented using spaces that are bounded by closed shapes, such as circles or ellipses for 2-dimensional Euclidean spaces, or spheres or polyhedrons for 3-dimensional Euclidean spaces. These shapes are called Venn-Euler (or simply Venn) diagrams. Belonging, nonbelonging, and overlaps between events and sets can be represented by these diagrams.

Example 4.4   A four-event Venn diagram

The Venn diagram in Figure 4.1 shows four events (or sets) A, B, C, and D that belong to a two-dimensional Euclidean sample space S. The event $C \subset B$ and $A \neq B$. Also, the events A and B have an overlap in the sample space S. Events A, B, and C are convex sets, whereas D is not. A convex set is defined to have the property that any straight line connecting any two elements of the set does not cross and go outside the set's boundaries. For $C \subset B$, the following inequality holds representing a monotonic behavior called *monotonousness*:

$$\mu_C(x) < \mu_B(x) \quad \text{for all } x \notin S \quad (4.16)$$

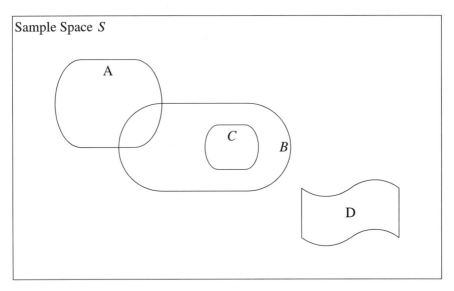

Figure 4.1 Crisp events or sets in two-dimensional Euclidean space.

4.2.2.6 Basic operations on sets

In this section, basic operations that can be used for sets and events are introduced. These operations are analogous to addition, subtraction, and multiplication in arithmetic calculations.

The *union* of events A and B is the set $A \cup B$ of all elements which belong to A or B or both. More events are called *collectively exhaustive* if the union of these events results in the sample space. The *intersection* of events A and B is the set $A \cap B$ of all elements which belong to both A and B. Events are called *mutually exclusive* if the occurrence of one event precludes the occurrence of other events. The difference of events A and B is the set $A - B$ of all elements which belong to A but which do not belong to B. The event that contains all the elements that do not belong to an event A, is called the complement of A, and is denoted $\overline{A}$. Table 4.1 shows additional rules based on the above basic rules.

The Cartesian product set of A and B is the set of all ordered pairs (a,b) such that $a \in A$ and $b \in B$. In mathematical notations, this can be expressed as

$$A \times B = \{(a,b) : a \in A \text{ and } b \in B\} \tag{4.17}$$

The cardinality of a set A is the number of elements in A and is denoted as $|A|$. For a given set A, the set of all subsets of A is called the *power set* of A, and is denoted P_A that has $2^{|A|}$ subsets in it.

Chapter four: Expressing and modeling expert opinions

Table 4.1 Other Operational Rules

Rule Type	Operations
Identity laws	$A \cup \emptyset = A$, $A \cap \emptyset = \emptyset$, $A \cup S = S$, $A \cap S = A$
Idempotent laws	$A \cup A = A$ and $A \cap A = A$
Complement laws	$A \cup \bar{A} = S$, $A \cup \bar{A} = \emptyset$, $\bar{\bar{A}} = A$, $\bar{S} = \emptyset$, $\bar{\emptyset} = S$
Commutative laws	$A \cup B = B \cup A$, $A \cap B = B \cap A$
Associative laws	$(A \cup B) \cup C = A \cup (B \cup C)$, $(A \cap B) \cap C = A \cap (B \cap C)$
Distributive laws	$(A \cup B) \cap C = (A \cap C) \cup (B \cap C)$
	$(A \cap B) \cup C = (A \cup C) \cap (B \cup C)$
de Morgan's law	$\overline{A \cup B} = \bar{A} \cap \bar{B}$, $\overline{(E_1 \cup E_2 \cup \cdots \cup E_n)} = \bar{E}_1 \cap \bar{E}_2 \cap \cdots \cap \bar{E}_n$
	$\overline{A \cap B} = \bar{A} \cup \bar{B}$, $\overline{(E_1 \cap E_2 \cap \cdots \cap E_n)} = \bar{E}_1 \cup \bar{E}_2 \cup \cdots \cup \bar{E}_n$
Combinations of laws	$\overline{(A \cup (B \cap C))} = \bar{A} \cap \overline{(B \cap C)} = (\bar{A} \cap \bar{B}) \cup (\bar{A} \cap \bar{C})$

Example 4.5 Power set and cardinality

For example, the following set A is used to determine its power set and cardinality:

$$A = \{1,2,3\} \quad (4.18)$$

The set A has the following power set:

$$P_A = \{\emptyset, \{1\}, \{2\}, \{3\}, \{1,2\}, \{1,3\}, \{2,3\}, \{1,2,3\}\} \quad (4.19)$$

These sets have the following respective cardinalities:

$$|A| = 3 \quad (4.20)$$

$$|P_A| = 8 \quad (4.21)$$

4.2.3 Fundamentals of fuzzy sets and operations

As stated earlier, the fundamental difference between classical (crisp or non-fuzzy) sets and fuzzy sets is that belonging and nonbelonging to sets are assumed to be without any form of uncertainty in the former and with uncertainty in the latter.

4.2.3.1 Membership (or characteristic) function

Let X be a universe, or a set of x values, and let A be a subset of X. Each element, x, is associated with a membership value to the subset A, $\mu A(x)$. For a fuzzy set, the membership function is given by

$$\mu_A : X \to [0,1] \tag{4.22}$$

as a mapping from the universe X to the interval of real values [0,1], where a value in this range means the grade of membership of each element x of X to the set A, i.e., the value of μ_A for each $x \in X$ can be viewed as a measure of the degree of compatibility of x with respect to the concept represented by A, where a value of 1 = the highest degree of compatibility and 0 = no compatibility. Fuzzy set analysis requires an analyst to manipulate the fuzzy set membership values; therefore a simplified notation for the membership function would facilitate this manipulation. According to a simplified membership notation, A is used to represent both a fuzzy set and its membership function. For example, $A(x)$ means the membership value of x to A.

For a fuzzy set A consisting of m discrete elements, the membership function can also be expressed as

$$A = \{ x_1 | \mu_A(x_1) , x_2 | \mu_A(x_2) , \ldots, x_m | \mu_A(x_m) \} \tag{4.23}$$

in which "=" should be interpreted as "is defined to be" and " | " = a delimiter. For continuous elements $x \in X$, the membership function of A can be expressed as

$$A = \{ x | \mu_A(x) \text{ or } A(x), \text{ for all } x \in X \} \tag{4.24}$$

in which the function $A(x)$ or $\mu_A(x)$ takes values in the range [0,1]. In the case of fuzzy sets, the boundaries of A are not sharp, and the membership of any x to A is fuzzy. The support of a fuzzy set is defined as all $x \in X$ such that $\mu_A(x) > 0$. A fuzzy set A is called to be a *subset* of or equal to a fuzzy set B, $A \subseteq B$, if and only if $\mu_A(x) \leq \mu_B(x)$ for all $x \in X$. A fuzzy set A is called to be equal to fuzzy set B, $A = B$, if and only if $\mu_A(x) = \mu_B(x)$ for all $x \in X$.

Fuzzy sets are able to express more realistically gradual transitions from membership to nonmembership to sets. Experts sometimes might find more meaningful expressions, although context dependent, to provide opinions using vague terms and words in natural languages, such as, likely, large, and poor quality. These vague terms cannot be modeled using crisp sets. Membership functions can be constructed subjectively based on experience and judgment through techniques identified in Chapters 3 and 6 and demonstrated in Chapter 7.

Chapter four: Expressing and modeling expert opinions

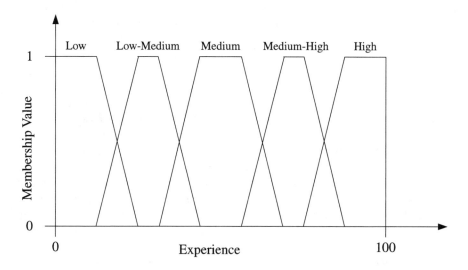

Figure 4.2 Experience levels as fuzzy sets.

Example 4.6 Fuzzy sets to represent experience

As an example of fuzzy sets, let X be the experience universe of an individual to perform some job, such that $x \in X$ can take a real value in the range from $x = 0$ meaning *absolutely no experience in performing the job* to $x = 100$ meaning *absolutely the highest level of experience*. The range 0 to 100 was selected for representation convenience. Other ranges could be used such as 0 to 1. Five levels of experience are shown in Figure 4.2 using linguistic descriptors, such as low, medium, and high experience. These experience classification are meaningful although vague. A fuzzy set representation offers a means of translating this vagueness into meaningful numeric expressions using membership functions.

Another method of presenting fuzzy sets that was used in the 1980s is based on dividing the range of experience into increments of 10. Therefore, a linguistic variable of the type "low or short experience," designated as A, can be expressed using the following illustrative fuzzy definition that does not correspond to Figure 4.2:

short experience, $A = \{ x_1 = 100 \mid \mu_A(x_1) = 0 , x_2 = 90 \mid \mu_A(x_2) = 0,$

$x_3 = 80 \mid \mu_A(x_3) = 0 , x_4 = 70 \mid \mu_A(x_4) = 0 , x_5 = 60 \mid \mu_A(x_5) = 0,$

$x_6 = 50 \mid \mu_A(x_6) = 0 , x_7 = 40 \mid \mu_A(x_7) = 0.1 , x_8 = 30 \mid \mu_A(x_8) = 0.5,$

$x_9 = 20 \mid \mu_A(x_9) = 0.7 , x_{10} = 10 \mid \mu_A(x_{10}) = 0.9 , x_{11} = 0 \mid \mu_A(x_{11}) = 1.0 \}$ (4.25)

This expression can be written in an abbreviated form by showing experience levels with only non-zero membership values as follows:

$$\text{short experience, } A = \{\,40\,|\,0.1,\ 30\,|\,0.5,\ 20\,|\,0.7,\ 10\,|\,0.9,\ 0\,|\,1.0\,\} \qquad (4.26)$$

The fuzziness in the definition of short experience is obvious from Equations 4.25 or 4.26 as opposed to a definition in the form of Equation 4.9. Based on the fuzzy definition of short experience, different grades of experience have different membership values to the fuzzy set, "short experience A." The membership values are decreasing as a function of increasing grade of experience. In this example, the values of x with nonzero membership values are 40, 30, 20, 10, and 0, and the corresponding membership values are 0.1, 0.5, 0.7, 0.9, and 1.0, respectively. Other values of x larger than 40 have zero membership values to the subset A. These membership values should be assigned based on subjective judgment with the help of experts and can be updated with more utilization of such linguistic measures in real life applications. If a crisp set were used in this example of defining short experience, the value of x would be 0 with a membership value of 1.0. Similarly, long experience, B, can be defined as

$$\text{long experience, } B = \{\,100\,|\,1,\ 90\,|\,0.9,\ 80\,|\,0.7,\ 70\,|\,0.2,\ 60\,|\,0.1\,\} \qquad (4.27)$$

It should be noted that Equations 4.25 – 4.27 show experience taking discrete values for convenience only since values between these discrete values have membership values that can be computed using interpolation between adjacent values. In order to use fuzzy sets in practical problems, some operational rules similar to those used in classical set theory (Table 4.1) need to be defined.

4.2.3.2 Alpha-cut sets

Fuzzy sets can be described effectively and in a manner that can facilitate the performance of fuzzy-set operations using an important concept of an α-*cut*. For a fuzzy set A defined on universe X and a number α in the unit interval of membership [0, 1], the α-cut of A, denoted by $^{\alpha}A$, is the *crisp* set that consists all elements of A with membership degrees in A greater than or equal to α. An α-cut of A as a crisp set can be expressed as

$$^{\alpha}A = \{x: A(x) \geq \alpha\} \qquad (4.28)$$

which means the crisp sets of x values such that $A(x)$ is greater or equal to α. For fuzzy sets of continuous real values and membership functions with one mode, i.e., unimodal functions, the resulting α-cut of A is an interval of real values. Such fuzzy sets and membership functions are common in engineering and sciences. A *strong* α-cut of A is defined as

Chapter four: Expressing and modeling expert opinions

$$^{\alpha+}A = \{x: A(x) > \alpha\} \quad (4.29)$$

The α-cut of A, with α = 1.0, is called the *core set* of the fuzzy set A. A fuzzy set with an empty core set is called a *subnormal fuzzy set* since the largest value of its membership function is less than one; otherwise the fuzzy set is called a *normal set*.

A set of nested $^{\alpha}A$ can be constructed by incrementally changing the value of α. A convenient representation of such nested sets at quartile α values is as follows:

$$\text{Nested } ^{\alpha}A = \{(\alpha_i, x_{Li}, x_{Ri})\} \text{ for } i = 1,2,3,4,5 \quad (4.30)$$

where the five sets of triplets correspond to α values of 1, 0.75, 0.50, 0.25, and 0+, respectively, and L and R refer to left and right, respectively. The α–cut of A, with α+ = 0+, is called the *support set* of the fuzzy set A. Other quartile levels can be termed *upper quartile set*, *mid quartile set*, and *lower quartile set*.

Example 4.7   α-cut of experience

Figure 4.3 shows a fuzzy-set representation or expression of medium experience based on the subjective assessment of an expert. The fuzzy set has a core set defined by the real range [40,60], and a support defined by [20,80]. The α-cuts of A are shown in Table 4.2 for all the quartile values and can be expressed as:

$$\text{Nested } ^{\alpha}A \text{ for Medium Experience} =$$
$$\{(1,40,60), (0.75,35,65), (0.5,30,70), (0.25,25,75), (0+,20,80)\} \quad (4.31)$$

4.2.3.3   Fuzzy Venn-Euler diagrams

Similar to crisp events, fuzzy events and sets can be represented using spaces that are bounded by closed shapes, such as circles with fuzzy boundaries showing the transitional stage from membership to nonmembership. Belonging, nonbelonging, and overlaps between events and sets can be represented by these diagrams. Figure 4.4 shows an example fuzzy event (or set) A with fuzzy boundaries. The various nested shapes can be considered similar to contours in topographical representations that correspond to the five-quartile α-cuts of A.

Table 4.2 α-cuts of Medium Experience as Provided in Figure 4.3

α	Left limit (x_L)	Right limit (x_R)	Name of set
1.	40	60	Core set
0.75	35	65	Upper quartile set
0.50	30	70	Mid quartile set
0.25	25	75	Lower quartile set
0.	20	80	Support set

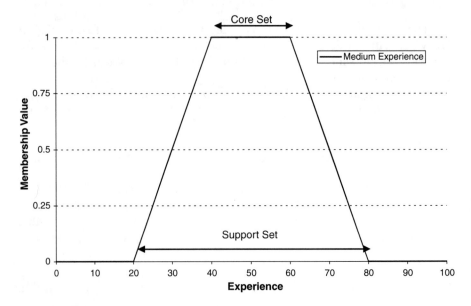

Figure 4.3 α-cut for medium experience.

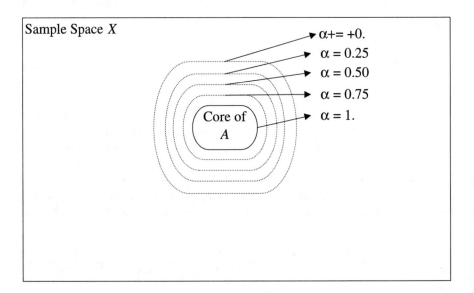

Figure 4.4 A fuzzy event.

Chapter four: Expressing and modeling expert opinions

4.2.3.4 Fuzzy numbers, intervals and arithmetic

A fuzzy number is defined as an approximate numerical value whose boundary is not sharp and can be characterized by a membership grade function. The membership function $\mu_A(x)$ for a fuzzy number A can be defined as given in Equation 4.22. An α-cut of a fuzzy number A can also be defined as provided by Equation 4.28. If the shape of the membership function is triangular, the fuzzy set is called a triangular fuzzy number. An α-cut of a fuzzy number (or set) A can be expressed as either by the five quartiles of Equation 4.30 or for an individual α as

$$^\alpha A = [x_L, x_R] = [a_L, a_R] \tag{4.32}$$

where L and R are the left and right x limits of the interval at α. A triangular fuzzy number A can be denoted as $A[a, a_m, b]$, where a = the lowest value of support, b = highest value of support, and a_m = is the middle value at the mode of the triangle. A trapezoidal fuzzy interval A can similarly be denoted as $A[a, a_{mL}, a_{mR}, b]$, where a = the lowest value of support, b = highest value of support, and a_{mL} and a_{mR} = are the left and right middle values at the mode range of the trapezoid. A triangular fuzzy number is commonly used to represent an approximate number, such as, the weight of a machine is approximately 200 pounds; whereas a trapezoidal fuzzy interval is an approximate interval, such as, the capacity of a machine is approximately 200 to 250 pounds. These examples of fuzzy numbers and fuzzy intervals are shown in Figure 4.5 with their crisp counterparts, where a crisp number can be represented as

$$A = a_m \tag{4.33}$$

A fuzzy number $A[a, a_m, b]$ can be represented as

$$A(x) = \begin{cases} \dfrac{x-a}{a_m - a} & \text{for } a \leq x \leq a_m \\ \dfrac{x-b}{a_m - b} & \text{for } a_m \leq x \leq b \\ 0 & \text{otherwise} \end{cases} \tag{4.34}$$

A crisp interval can be represented as

$$A = [a_{m1}, a_{m2}] \tag{4.35}$$

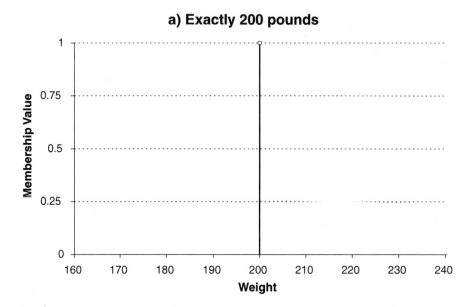

Figure 4.5a Crisp number.

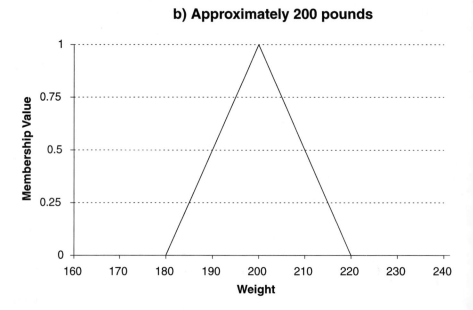

Figure 4.5b Fuzzy numbers.

Chapter four: Expressing and modeling expert opinions 139

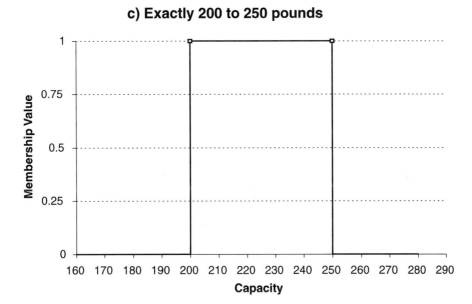

Figure 4.5c Crisp interval.

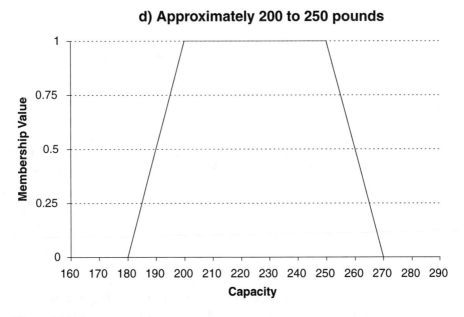

Figure 4.5d Fuzzy interval.

A fuzzy interval $A[a, a_{mL}, a_{mR}, b]$ can be represented as

$$A(x) = \begin{cases} \dfrac{x-a}{a_{m_1} - a} & \text{for } a \leq x \leq a_{m_1} \\[6pt] \dfrac{x-b}{a_{m_2} - b} & \text{for } a_{m_2} \leq x \leq b \\[6pt] 0 & \text{otherwise} \end{cases} \qquad (4.36)$$

For two fuzzy numbers or intervals A and B, let $^{\alpha}A = [a_L, a_R]$ and $^{\alpha}B = [b_L, b_R]$, where a and b are real numbers on the left (L) and right (R) ends of the ranges, and $\alpha \in [0,1]$ The fuzzy arithmetic of addition, subtraction, multiplication, and division, respectively, can be defined as follows (Kaufmann and Gupta, 1985) based on interval-valued arithmetic (Moore, 1966 and 1979):

$$^{\alpha}A + {}^{\alpha}B = [a_L, a_R] + [b_L, b_R] = [a_L + b_L, a_R + b_R] \qquad (4.37)$$

$$^{\alpha}A - {}^{\alpha}B = [a_L, a_R] - [b_L, b_R] = [a_L - b_R, a_R - b_L] \qquad (4.38)$$

$$^{\alpha}A \times {}^{\alpha}B = [a_L, a_R] \times [b_L, b_R] = [\min(a_L b_L, a_L b_R, a_R b_L, a_R b_R), \qquad (4.39)$$
$$\max(a_L b_L, a_L b_R, a_R b_L, a_R b_R)]$$

$$^{\alpha}A / {}^{\alpha}B = [a_L, a_R] / [b_L, b_R] = [\min(a_L/b_L, a_L/b_R, a_R/b_L, a_R/b_R), \qquad (4.40)$$
$$\max(a_L/b_L, a_L/b_R, a_R/b_L, a_R/b_R)]$$

Equation 4.40 requires that $0 \notin [b_L, b_R]$. The above equations can be used to propagate interval input into input-output models to obtain interval outputs using methods such as the *vertex method* (Dong and Wong, 1986a, 1986b, and 1986c).

In order to employ fuzzy arithmetic in numerical methods, the fuzzy subtraction and division of Eqations 4.38 and 4.40, respectively, should be revised to the constraint type (Ayyub and Chao, 1998) as defined by Klir and Cooper (1996). For example, the definition of fuzzy division for a fuzzy number by another fuzzy number of the same magnitude can be different than the fuzzy division of a fuzzy number by itself. Such a difference for $^{\alpha}A / {}^{\alpha}A$ can be provided for $^{\alpha}A = [a_L, a_R]$ with $0 \notin [a_L, a_R]$, and for all $x \in {}^{\alpha}A$ and $y \in {}^{\alpha}A$ as follows:

1. For nonrestricted x and y, the unconstraint fuzzy division based on Equation 4.40 can be expressed as:

Chapter four: Expressing and modeling expert opinions 141

$$\frac{^{\alpha}A(x)}{^{\alpha}A(y)} = \frac{[a_L, a_R]}{[a_L, a_R]} = [\min(a_L/a_L, a_L/a_R, a_R/a_L, a_R/a_R) \quad (4.41)$$
$$\max(a_L/a_L, a_L/a_R, a_R/a_L, a_R/a_R)]$$

2. For a constraint case where $x = y$, the fuzzy division is given by

$$\left.\frac{^{\alpha}A(x)}{^{\alpha}A(y)}\right|_{x=y} = \frac{[a_L, a_R]}{[a_L, a_R]} = 1 \quad (4.42)$$

For fuzzy subtraction, a similar definition for $^{\alpha}A - {^{\alpha}A}$ can be given for all $x \in {^{\alpha}A}$ and $y \in {^{\alpha}A}$ as follows:

1. For nonrestricted x and y, the unconstraint fuzzy subtraction based on Equation 4.38 is given by

$$^{\alpha}A(x) - {^{\alpha}A(y)} = [a_L, a_R] - [a_L, a_R] = [a_L - a_R, a_R - a_L] \quad (4.43)$$

2. For a restricted case where $x = y$, the constraint fuzzy subtraction is

$$[^{\alpha}A(x) - {^{\alpha}A(y)}]_{x=y} = [a_L, a_R] - [a_L, a_R] = [a_L - a_L, a_R - a_R] = 0 \quad (4.44)$$

Constraint fuzzy arithmetic is most needed for performing numerical manipulation in solving problems, for example, a system of linear equations with fuzzy coefficients and other numerical problems. Similar definitions can be provided for fuzzy addition and multiplication. In addition, the constraint is not limited to $x = y$. The concept can be extended to any constraint, such as, equalities of the type $x + y = 100$ and $x^2 + y^2 = 1$, or inequalities of the type $x < y$ and $x^2 + y^2 \leq 1$. The inequality constraints require the use of union operations to deal with numerical answers that can be produced by several x and y combinations, i.e., lack of uniqueness or mapping from many to one. Additional information on constraint arithmetic is provided by Klir and Cooper (1996).

Fuzzy arithmetic can be used to develop methods for aggregating expert opinions that are expressed in linguistic or approximate terms. This aggregation procedure retains uncertainties in the underlying opinions by obtaining a fuzzy combined opinion.

Example 4.8 Additional operations on fuzzy sets

The following two fuzzy numbers are used to perform a series of arithmetic operations as provided below for demonstration purposes:

$$A = [1,2,3], \text{ a triangular fuzzy number}$$
$$B = [2,4,6], \text{ a triangular fuzzy number}$$
$$A + B = B + A = [3,6,9], \text{ a triangular fuzzy number}$$
$$A - B = [-5,-2,1], \text{ a triangular fuzzy number}$$

The multiplication and division do not produce triangular numbers, and they need to be evaluated using α-cuts. For example at α = 0, 0.5, and 1, the intervals for the product and division are

$$\text{At } \alpha = 0, A \times B = [2,18]$$
$$\text{At } \alpha = 0.5, A \times B = [4.5,12.5]$$
$$\text{At } \alpha = 1, A \times B = [8,8]$$
$$\text{At } \alpha = 0, B/A = [2/3,6]$$
$$\text{At } \alpha = 0.5, B/A = [1.2,10/3]$$
$$\text{At } \alpha = 1, B/A = [2,2]$$

Figure 4.6 shows graphically the results of addition, subtraction, multiplication, division, constraint addition, and constraint multiplication at various α-cuts.

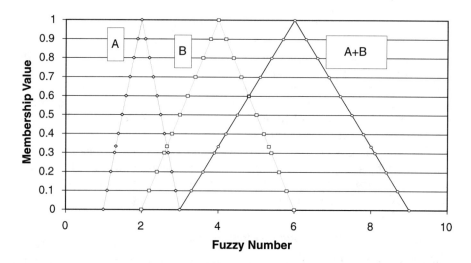

Figure 4.6a Example fuzzy arithmetic of addition.

Chapter four: Expressing and modeling expert opinions 143

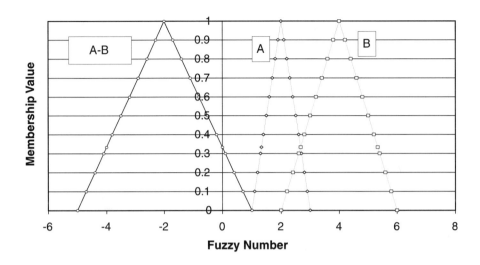

Figure 4.6b Example fuzzy arithmetic of subtraction.

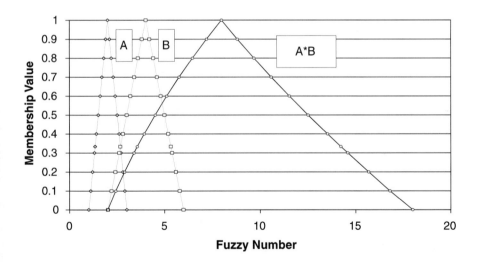

Figure 4.6c Example fuzzy arithmetic of multiplication.

4.2.3.5 *Operations on fuzzy sets*

In this section, basic operations that can be used for fuzzy sets and events are introduced. These operations are defined in an analogous form to the corresponding operations of crisp sets in Section 4.2.2.6.

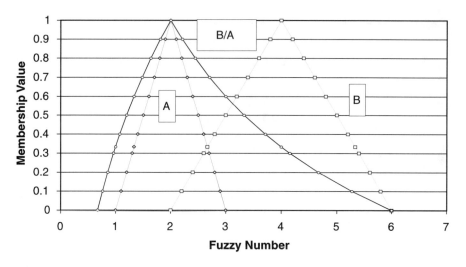

Figure 4.6d Example fuzzy arithmetic of division.

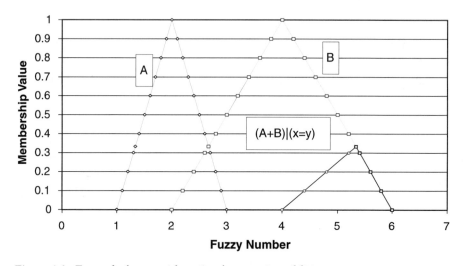

Figure 4.6e Example fuzzy arithmetic of constraint addition.

The *union* of sets $A \subset X$ and $B \subset X$ is the set $A \cup B \subset X$ which corresponds to the connective "or," and its membership function is given by

$$\mu_{A \cup B}(x) = \max\;[\mu_A(x),\; \mu_B(x)\;] \qquad (4.45)$$

This equation can be generalized to obtain what is called the *triangular conorms*, or for short the *t-conorms*, such as the *Yager-class* of fuzzy unions, provided by

Chapter four: Expressing and modeling expert opinions

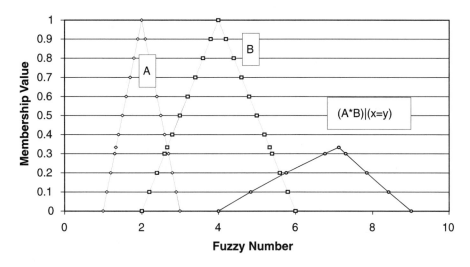

Figure 4.6f Example fuzzy arithmetic of constraint multiplication.

$$\mu_{A \cup B}(x) = \min\left[1, \sqrt[\beta]{(\mu_A(x))^\beta + (\mu_B(x))^\beta}\right] \quad (4.46)$$

where $\beta \in (0, \infty)$ is called the *intensity factor*. Equation 4.46 reduces to Equation 4.45 as $\beta \to \infty$. The union based on Equation 4.46 depends on β, and can take any value in the following range with lower and upper limits that correspond to $\beta \to \infty$ and $\beta \to 0$, respectively:

$$\max(\mu_A(x), \mu_A(x)) \le \mu_{A \cup B}(x) \le \begin{cases} \mu_A(x) & \text{if } \mu_B(x) = 0 \\ \mu_B(x) & \text{if } \mu_A(x) = 0 \\ 1 & \text{otherwise} \end{cases} \quad (4.47)$$

The *intersection* of sets $A \subset X$ and $B \subset X$ is the set $A \cap B \subset X$ which corresponds to the connective "and," and its membership function is given by

$$\mu_{A \cap B}(x) = \min\left[(\mu_A(x), (\mu_B(x))\right] \quad (4.48)$$

This equation can be generalized to obtain what is called the *triangular norms*, or for short the *t-norms*, such as the *Yager-class* of fuzzy intersections, provided by

$$\mu_{A \cap B}(x) = 1 - \min\left[1, \sqrt[\beta]{(1-\mu_A(x))^\beta + (1-\mu_B(x))^\beta}\right] \quad (4.49)$$

where $\beta \in (0,\infty)$ is called the *intensity factor*. Equation 4.49 reduces to Equation 4.48 as $\beta \to \infty$. The intersection based on Equation 4.49 depends on β, and can take any value in the following range with lower and upper limits that correspond to $\beta \to 0$ and $\beta \to \infty$, respectively:

$$\left. \begin{array}{ll} \mu_A(x) & \text{if } \mu_B(x) = 1 \\ \mu_B(x) & \text{if } \mu_A(x) = 1 \\ 0 & \text{otherwise} \end{array} \right\} \le \mu_{A \cap B}(x) \le \min(\mu_A(x), \mu_B(x)) \quad (4.50)$$

The *difference* between events A and B is the set $A - B$ of all elements that belong to A but do not belong to B. The difference is mathematically expressed as

$$\mu_{A-B}(x) = \begin{cases} \mu_A(x) & \text{if } \mu_B(x) = 0 \\ 0 & \text{if } \mu_B(x) \ne 0 \end{cases} \quad (4.51)$$

The membership function of the complement $\overline{A}$ of a fuzzy set A is given by

$$\mu_{\overline{A}}(x) = 1 - \mu_A(x) \quad (4.52)$$

This equation can be generalized to obtain the *Yager-class* of fuzzy complements as follows:

$$\mu_{\overline{A}_\beta}(x) = \sqrt[\beta]{1 - (\mu_A(x))^\beta} \quad (4.53)$$

where $\beta \in (0,\infty)$ is called the *intensity factor*. Equation 4.53 reduces to Equation 4.52 for $\beta = 1$. The definition of the complement can be generalized as provided, for example, by the *Sugeno-class* of fuzzy complements as follows:

$$\mu_{\overline{A}_\beta}(x) = \frac{1 - \mu_A(x)}{1 + \beta \mu_A(x)} \quad (4.54)$$

where $\beta \in (0,\infty)$ is called the *intensity factor*. Equation 4.54 reduces to Equation 4.52 as $\beta = 0$. The selection of a β value in the generalized definitions of fuzzy set operations requires the use of experts to calibrate these

Chapter four: Expressing and modeling expert opinions

operations based on the context of use and application. The generalized definitions according to Equations 4.46, 4.49, 4.53, and 4.54 contain *softer* definitions of union, intersection and complement, whereas the definitions of the union, intersection, and complement of Equations 4.45, 4.48, and 4.52 offer *hard* definitions. The generalized unions and intersections for two fuzzy sets produce membership values as shown in Figure 4.7.

Figure 4.7 shows *averaging*, also called *aggregation*, operations that span the gap between the unions and intersections (Klir and Folger, 1988; Klir and Wierman, 1999). An example of generalized aggregation is the *generalized means* and is defined by

$$\mu_{A \triangledown B}(x) = \frac{1}{2} \sqrt[\beta]{(\mu_A(x))^\beta + (\mu_B(x))^\beta} \quad (4.55)$$

where $\triangledown$ is the *generalized means* aggregation operator, and β is an *intensity factor* whose range is the set of all real numbers excluding 0. Equation 4.55 becomes the geometric mean as $\beta \to 0$, and becomes

$$\mu_{A \triangledown B}(x) = \sqrt{\mu_A(x)\mu_B(x)} \quad (4.56)$$

Equation 4.55 converges to the min and max operations as $\beta \to -\infty$ and $\beta \to \infty$, respectively. Figure 4.7 shows this convergence.

Other generalized operations and additional information on this subject are provided by Klir and Folger (1988), Klir and Wierman (1999), Yager (1980a), Schweizer and Sklar (1983), Frank (1979), Dubois and Prade (1980), and Dombi (1982).

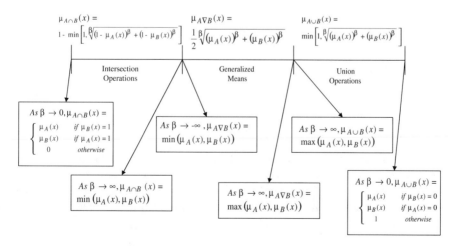

Figure 4.7 Generalized fuzzy operations.

Example 4.9 Operations on fuzzy experience levels

Two experts provided the following assessments of long experience using fuzzy sets B and C:

$$B = \{\text{ 100 | 1. , 90 | 0.9 , 80 | 0.7 , 70 | 0.2 , 60 | 0.1 }\} \quad (4.57)$$

long experience,

$$C = \{\text{ 100 | 1. , 90 | 0.8 , 80 | 0.6 , 70 | 0.4 , 60 | 0.2 }\} \quad (4.58)$$

The union and intersection of these two sets can be computed according to Equations 4.45 and 4.48, respectively, to obtain the following:

$$B \cup C = \{100 \mid 1. , 90 \mid 0.9 , 80 \mid 0.7 , 70 \mid 0.4 , 60 \mid 0.2 \} \quad (4.59)$$

$$B \cap C = \{100 \mid 1. , 90 \mid 0.8 , 80 \mid 0.6 , 70 \mid 0.2 , 60 \mid 0.1 \} \quad (4.60)$$

The above definitions of the union and intersection of fuzzy sets are the hard definitions. The difference of the two sets is the empty set, and can be stated as

$$B - C = \emptyset \quad (4.61)$$

The complement of B according to Equation 4.52 is given by

$$\mu_{\bar{B}}(x) = \{90 \mid 0.1 , 80 \mid 0.3 , 70 \mid 0.8 , 60 \mid 0.9 , 50 \mid 1. , 40 \mid 1. ,$$
$$30 \mid 1. , 20 \mid 1. , 10 \mid 1. , 0 \mid 1. \} \quad (4.62)$$

The α-cut of B at $\alpha = 0.7$ according to Equation 4.28 is given by an interval of real values as follows:

$$B_{0.7} = [80 , 100] \quad (4.63)$$

Example 4.10 Additional operations on fuzzy sets

Figure 4.8 shows example fuzzy operations on two fuzzy events A and B. The intersection and union are shown using the min and max rules of Equations 4.48 and 4.45, respectively. The complement is also shown based on Equation 4.52. The last two figures show the unique properties of fuzzy sets of $A \cup \bar{A} \neq S$ and $A \cap \bar{A} \neq \emptyset$.

Chapter four: Expressing and modeling expert opinions 149

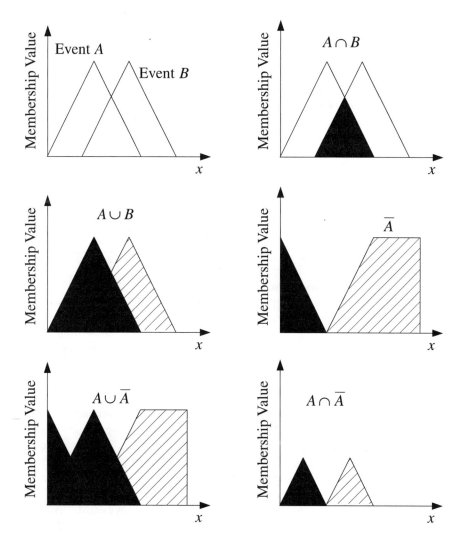

Figure 4.8 Example fuzzy operations.

Example 4.11 Failure definition for reliability and safety studies of structures

Classical structural reliability assessment techniques are based on precise and crisp definitions of failure and survival of a structure in meeting a set of strength, functional, and serviceability criteria. Consider the following performance function:

where $X_1, X_2, \ldots, X_n$ = basic random variables, and Z = performance measure or safety margin as the difference between structural strength as a response (R) to loads and the applied loads (L) (i.e., $Z = R - L$). Both R and L are functions of the basic random variables. Equation 4.64 is defined such that failure occurs where $Z < 0$, survival occurs where $Z > 0$, and the limit state equation is defined as $Z = 0$. The probability of failure can then be determined by solving the following integral:

$$P_f = \iint \cdots \int f_{\underline{X}}(x_1, x_2, \ldots, x_n)\, dx_1 dx_2 \ldots dx_n \qquad (4.65)$$

where $f_{\underline{X}}$ is the joint probability density function of $\{X_1, X_2, \ldots, X_n\}$, and the integration is performed over the region where $Z < 0$ (Ayyub and McCuen, 1997; Ayyub and Haldar, 1984; White and Ayyub, 1985).

The model for crisp failure consists of two basic, mutually exclusive events, i.e., complete survival and complete failure. The transition from one to another is abrupt rather than continuous. This model is illustrated in Figure 4.9, where R_f is the structural response at the limiting state for a selected design criterion. If the structural response R is smaller than R_f, i.e., $R < R_f$, the complete survival state exists and it is mapped to the zero failure level ($\alpha = 0$). If the structural response R is larger than R_f, i.e., $R > R_f$, the complete failure state occurs and it is mapped to $\alpha = 1$. The limit state is defined where $R = R_f$.

The fuzzy failure model is illustrated by introducing a subjective failure level index α as shown in Figure 4.10, where R_L and R_R are the left (lower) bound and right (upper) bound of structural response for the region of transitional or partial failure, respectively. The complete survival state is defined where $R \leq R_L$, the response in the range $(R_L < R < R_U)$ is the transitional state, and the response $(R \geq R_U)$ is the complete failure state. In Figure 4.10, the structural response is mapped to the failure-level scale to model some performance event as follows:

$$\text{Performance event } A: R \rightarrow A = \{\alpha : \alpha \in [0,1]\} \qquad (4.66)$$

where 0 = failure level for complete survival, 1 = failure level for complete failure, and $[0, 1]$ = all real values in the range of 0 to 1 for all failure levels.

The index α can also be interpreted as a measure of degree of belief in the occurrence of some performance condition. In this case, $\alpha = 0$ is interpreted as no belief of the occurrence of an event, and the $\alpha = 1$ means absolute belief of the occurrence of the event.

A mathematical model for structural reliability assessment that includes both ambiguity and vagueness types of uncertainty was suggested by Alvi

Chapter four: Expressing and modeling expert opinions

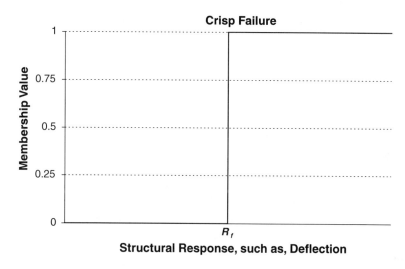

Figure 4.9 A crisp failure definition.

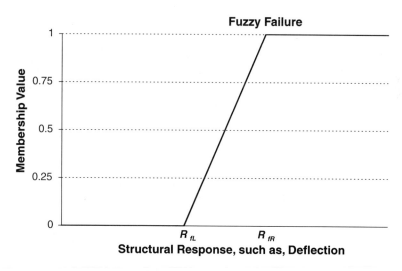

Figure 4.10 A fuzzy failure definition.

and Ayyub (1990) and Ayyub and Lai (1992 and 1994). The model results in the likelihood of failure over a damage spectrum. Since the structural reliability assessment is based on the fuzzy failure model, the probability of failure, in this case, is a function of α. Figure 4.11 shows various grades of failures expressed as fuzzy events as used by Ayyub and Lai (1992) to assess the structural reliability of ship hull girders.

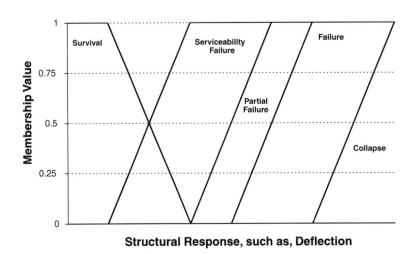

Figure 4.11 Fuzzy failure definitions for structural reliability assessment.

4.2.3.6 *Fuzzy relations*

A relation between two or more sets is defined as an expression of association, interrelationship, interconnection, or interaction among these sets. The expression can be made in a crisp format indicating the presence or absence of such a relation, or it can be made in a fuzzy format indicating the degree of belief in the presence or absence of the relation. A degree of belief in a relationship R between $x \in X$ and $y \in Y$ is commonly expressed using $\mu_R(x,y) \in [0,1]$. If $\mu_R(x,y) \in \{0,1\}$, i.e., $\mu_R(x,y)$ can take one of two values 0 or 1, then R is considered to be a crisp relation; whereas if $\mu_R(x,y) \in [0,1]$, i.e., $\mu_R(x,y)$ can take any real value in the range [0,1], then R is considered to be a fuzzy relation.

A fuzzy relation R between two sets $A \subset X$ and $B \subset Y$ can be defined on the Cartesian product of A and B as the set of all ordered pairs (a,b) such that $a \in A$ and $b \in B$, for discrete A and B. In mathematical notations, the Cartesian product is expressed by the following space:

$$A \times B = \{(a,b): a \in A \text{ and } b \in B\} \tag{4.67}$$

For continuous A and B, the Cartesian product can be shown as a 2-dimensional x-y space over the ranges of A and B. The membership function for a fuzzy relation can be defined as

$$\mu_R(a,b) = \min\,[\mu_A(a)\,,\,\mu_B(b)] \tag{4.68}$$

Chapter four: Expressing and modeling expert opinions

For discrete A and B, relations are usually expressed in a matrix form as

$$R = A \times B = \begin{array}{c|cccc} & b_1 & b_2 & \cdots & b_m \\ \hline a_1 & \mu_R(a_1,b_1) & \mu_R(a_1,b_2) & \cdots & \mu_R(a_1,b_m) \\ a_2 & \mu_R(a_2,b_1) & \mu_R(a_2,b_2) & \cdots & \mu_R(a_2,b_m) \\ \vdots & \vdots & \vdots & \vdots & \vdots \\ a_n & \mu_R(a_n,b_1) & \mu_R(a_n,b_2) & \cdots & \mu_R(a_n,b_m) \end{array} \quad (4.69)$$

in which $A = \{a_1, a_2, ..., a_n\}$; $B = \{b_1, b_2, ..., b_m\}$; and $\mu_R(a_i,b_j)$ = support or membership value for the ordered pair (a_i,b_j) that is a measure of association between a_i and b_j. The membership value can be determined based on judgment or can be computed as the minimum value of the membership values $\mu_A(a_i)$ and $\mu_B(b_j)$.

A fuzzy relation can be expressed in a conditional form. For example, the relation R can be defined as

'if experience of workers on a production line is short, then the quality of the product is medium.

Defining "short experience" and a "medium" product quality, respectively, as

$$\text{short experience,}$$
$$A = \{\, 40 \mid 0.1\,,\, 30 \mid 0.5\,,\, 20 \mid 0.7\,,\, 10 \mid 0.9\,,\, 0 \mid 1.0\,\} \quad (4.70)$$

$$\text{medium quality} =$$
$$\{\, 70 \mid 0.2\,,\, 60 \mid 0.7\,,\, 50 \mid 1.\,,\, 40 \mid 0.7\,,\, 30 \mid 0.2\,\} \quad (4.71)$$

then, the fuzzy relation R can be computed as

		short experience				
		40	30	20	10	0
	70	0.1	0.2	0.2	0.2	0.2
medium	60	0.1	0.5	0.7	0.7	0.7
R = product	50	0.1	0.5	0.7	0.9	1.0
quality	40	0.1	0.5	0.7	0.7	0.7
	30	0.1	0.2	0.2	0.2	0.2

(4.72)

Note that the fuzzy sets "short experience" and "medium product quality" are from two different universes, namely, "experience" and "quality," respectively. The membership values of the first row in Equation 4.72 were evaluated as follows:

$$\mu_R(70, 40) = \min(0.2, 0.1) = 0.1$$
$$\mu_R(70, 30) = \min(0.2, 0.5) = 0.2$$
$$\mu_R(70, 20) = \min(0.2, 0.7) = 0.2 \quad (4.73)$$
$$\mu_R(70, 10) = \min(0.2, 0.9) = 0.2$$
$$\mu_R(70, 0) = \min(0.2, 1.0) = 0.2$$

The union of two relations, say R and S, is denoted by $R \cup S$ and has the following membership function:

$$\mu_{R \cup S}(x,y) = \max[\mu_R(x,y), \mu_S(x,y)] \quad (4.74)$$

where both relations R and S are defined on the Cartesian product space $X \times Y$. On the other hand, the intersection of two fuzzy relations, $R \cap S$, has the following membership function:

$$\mu_{R \cap S}(x,y) = \min[\mu_R(x,y), \mu_S(x,y)] \quad (4.75)$$

These "hard" definitions of the union and intersection of two fuzzy relations can be generalized to perform the union and intersection of several relations using the "max" and "min" operators, respectively. The hard definitions of these operations can also be generalized using the Yager-classes, similar to the union and intersection of fuzzy sets.

Other operations on fuzzy relations include their complement and composition. The complement $\bar{R}$ of fuzzy relation R has the following membership function:

$$\mu_{\bar{R}}(x,y) = 1 - \mu_R(x,y) \quad (4.76)$$

The composition of a fuzzy relation R defined on the universe $X \times Y$ and a fuzzy relation S defined on the universe $Y \times Z$ is the fuzzy relation $R \circ S$ defined on the universe $X \times Z$. The membership function for this fuzzy relation is given by

$$\mu_{R \circ S}(x,z) = \max_{\text{all } y \in Y} \{\min[\mu_R(x,y), \mu_S(y,z)]\} \quad (4.77)$$

An interesting case of fuzzy composition is the composition of a fuzzy subset A defined on the universe X with a relation R defined on the universe $X \times Y$. The result is a fuzzy subset B defined on the universe Y with a membership function given by

$$\mu_{A \circ R}(y) = \max_{\text{all } x \in X} \{\min[\mu_A(x), \mu_R(x,y)]\} \quad (4.78)$$

A common application of the above operations of fuzzy relations is in constructing an approximate logic based on conditional statements of the following type: "if A_1 then B_1 else if A_2 then B_2 ... else if A_n then B_n." This statement can be modeled using the operations of fuzzy sets and relations in the following form:

$$(A_1 \times B_1) \cup (A_2 \times B_2) \ldots \cup (A_n \times B_n) \tag{4.79}$$

where $A_1, A_2, \ldots, A_n$ and $B_1, B_2, \ldots, B_n$ are fuzzy sets. Equation 4.79 is used for developing controllers based on fuzzy logic.

Additional information about the above operations and other operations with examples are provided by Kaufman (1975), Kaufman and Gupta (1985), Klir and Folger (1988), and Hassan et al. (1992), Hassan and Ayyub (1993a, 1993b, 1994, 1997), and Ayyub and Hassan (1992a, 1992b, 1992c).

4.2.3.7 Fuzzy functions

Although the concept of fuzzy functions was explored by Sugeno (1974, 1977), and Wang and Klir (1992) by developing fuzzy integral within the framework of fuzzy measure theory, it was not presented in a format that met the needs of engineers and scientists or, for our purposes in this chapter, to express expert opinions. Engineers and scientists are commonly interested in developing relationships among underlying variables for an engineering system. These relationships are based on the underlying physics of modeled problems or system behavior, such as economic forecasts, power consumption forecasting, and extreme loads on a structure.

In this section, a format for presenting these subjective relationships is proposed, although the mathematics needed to manipulate these functions are not available yet. These manipulations can be developed using the extension principle based on fuzzy numbers, fuzzy intervals, and fuzzy arithmetic.

In engineering and science, we are interested in cause-effect relationships expressed as

$$Y = f(x) \tag{4.80}$$

where Y is the value of the criterion variable, also called the dependent variable; x is the predictor variable, also called the independent variable; and f is the functional relationship. Using the concept of α-cuts, a fuzzy function can be expressed using the following triplet functions:

$$Y = (^{\alpha T}f(x), {}^{\alpha 1}f(x), {}^{\alpha B}f(x)) \tag{4.81}$$

where $^{\alpha T}f(x)$ is the left or top α-cut function at $\alpha = 0$, $^{\alpha B}f(x)$ is the right or bottom α-cut function at $\alpha = 0$, and $^{\alpha 1}f(x)$ is the middle α-cut function at $\alpha = 1$. Other α-cut functions can be developed using linear interpolation as follows:

$$^{\alpha}f(x) = \begin{cases} ^{\alpha B}f(x) + a\left(^{\alpha 1}f(x) - ^{\alpha B}f(x)\right) & \text{for } ^{\alpha B}f(x) \le ^{\alpha}f(x) - ^{\alpha 1}f(x) \\ ^{\alpha 1}f(x) + a\left(^{\alpha T}f(x) - ^{\alpha 1}f(x)\right) & \text{for } ^{\alpha 1}f(x) \le ^{\alpha}f(x) - ^{\alpha T}f(x) \end{cases} \quad (4.82)$$

Fuzzy functions can be used to extrapolate empirical functions to regions beyond data availability, for example for developing forecasting models. Functional operations such as derivatives, integrals, roots, maximums, and minimums can be defined using the α-cut concepts. These computations can be performed using numerical techniques performed on the function values at the α-cuts. Ayyub and McCuen (1996) describe commonly used numerical methods with practical examples.

Example 4.12 Forecasting power needs using a fuzzy function

The power needs of a city can be forecasted to help city planners make zoning and power-plant construction decisions. Figure 4.12 shows an empirical power consumption trend over time and a subjectively developed forecast of the city's needs. The forecasted segment of the curve is provided for demonstration purposes. The following fuzzy function is used to develop Figure 4.12:

$$Y = \begin{cases} \alpha^T f(x) = 233 + 5\sqrt{5(\text{Year} - 2000)} \\ \alpha^1 f(x) = 233 + 5\sqrt{(\text{Year} - 2000)} \\ \alpha^B f(x) = 233 + 5\sqrt{(\text{Year} - 2000)/5} \end{cases} \quad (4.83)$$

where Year is in a four-digit number format. The figure shows the empirical data, and three functions that correspond to middle, and top and bottom at α-cuts of 0 and 1, respectively.

4.2.4 Fundamentals of rough sets

4.2.4.1 Rough set definitions

Rough sets were introduced by Pawlak (1991) and are described with examples by Pal and Skowron (1999). Rough sets provide the means to represent imprecise classical (crisp) sets. The representation is based on a partition of the universal space involved that should be constructed to facilitate this representation. Each rough set represents a given crisp set by two subsets of the partition, a lower approximation and an upper approximation. The lower approximation consists of all subsets of the partition that are included in the crisp set represented, whereas the upper approximation consists of all subsets that overlap with the set represented. Figure 4.13 shows a crisp set

Chapter four: Expressing and modeling expert opinions

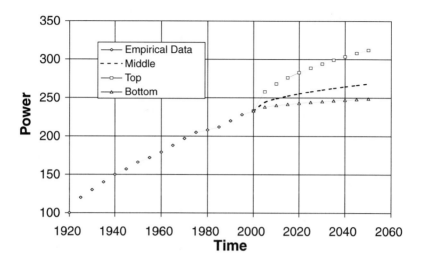

Figure 4.12 A fuzzy function for forecasting power needs.

A that belongs to a universal space S. The universal space is partitioned using the grid shown in Figure 4.13. The lower and upper approximations of A are shown in the figure. These two subsets of approximations constitute the rough set approximation of A.

Fuzzy sets are different from rough sets. The former represents vagueness of a quantity, such as obtaining linguistic quantities from experts, whereas, the latter represents coarseness as an approximation of a crisp set. Since both types can be relevant in some applications, they can be combined either as *fuzzy rough sets* or *rough fuzzy sets*. Fuzzy rough sets are rough sets that are based on fuzzy partitions, whereas rough fuzzy sets are rough-set approximations of fuzzy sets based on crisp partitions.

4.2.4.2 Rough set operations

Rough sets can be manipulated using operations of unions, intersection and complements. Figure 4.13 shows a rough set approximation of the crisp set A. The lower and upper rough (R) set approximation of A can be written as

$$\underline{R}(A) \subseteq A \subseteq \overline{R}(A) \tag{4.84}$$

where $\underline{R}(A)$ and $\overline{R}(A)$ are the lower and upper approximations of A. The subset $\underline{R}(A)$ includes all the sets of the universal space (S) that are contained in A. The subset $\overline{R}(A)$ includes all the sets of the universal space that contain any part of A. Based on this definition, the following operations can be provided for two crisp sets A and B (Pal and Skowron 1999):

a) Lower Approximation

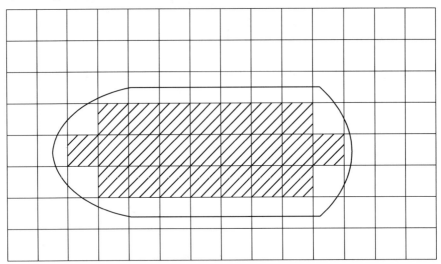

b) Upper Approximation

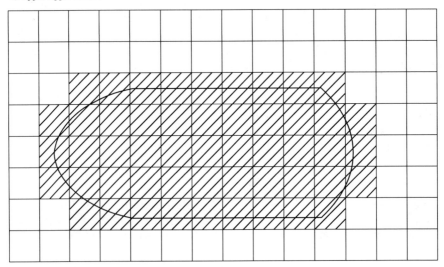

Figure 4.13 Rough set approximations.

$$\underline{R}(\emptyset) = \emptyset = R(\emptyset) \qquad (4.85)$$

$$\underline{R}(S) = S = R(S) \qquad (4.86)$$

$$\overline{R}(A \cup B) = \overline{R}(A) \cup \overline{R}(B) \qquad (4.87)$$

Chapter four: Expressing and modeling expert opinions

$$\underline{R}(A \cap B) = \underline{R}(A) \cap \underline{R}(B) \tag{4.88}$$

$$A \subseteq B \implies \underline{R}(A) \subseteq \underline{R}(B) \text{ and } \overline{R}(A) \subseteq \overline{R}(B) \tag{4.89}$$

$$\underline{R}(A \cup B) \supseteq \underline{R}(A) \cup \underline{R}(B) \tag{4.90}$$

$$\overline{R}(A \cap B) \subseteq \overline{R}(A) \cap \overline{R}(B) \tag{4.91}$$

$$\overline{R}(A) = \overline{\underline{R}(A)} \tag{4.92}$$

$$\underline{R}(A) = \underline{\overline{R}(A)} \tag{4.93}$$

$$\underline{R}(\underline{R}(A)) = \overline{R}(\underline{R}(A)) = \underline{R}(A) \tag{4.94}$$

$$\overline{R}(\overline{R}(A)) = \underline{R}(\overline{R}(A)) = \overline{R}(A) \tag{4.95}$$

A measure of the accuracy of an approximation (δ) can be expressed as

$$\delta = \frac{|\underline{R}(A)|}{|\overline{R}(A)|} \tag{4.96}$$

where $|\underline{R}(A)|$ is the cardinality of $\underline{R}(A)$, i.e., the number of elements in $\underline{R}(A)$. The range of δ is [0,1].

4.2.4.3 Rough membership functions

In classical sets, the concept of a membership function was introduced as provided by Equations 4.9 and 4.10. For each element, the membership function in the case of crisp sets takes on either 1 or 0 that correspond to belonging and not belonging of the element to a set A of interest. The membership function for rough sets has a different meaning than in the case of crisp sets. The *rough membership function* can be interpreted as the conditional probability that an element a belongs to the set A given the information represented by $\underline{R}(A)$ and $\overline{R}(A)$.

Example 4.13 Rough sets to represent information systems

This example deals with a quality assurance department in a factory that inspects all hand-made products produced by the factory. For each item produced, the inspectors at the department record who made it and the result of the inspection as either acceptable or unacceptable. Skilled workers used to produce these items have varying levels of experience in number of months at the job and number of items that each has produced. The data

Table 4.3 Inspection Data for Quality

Item Number	Experience (Number of Months)	Experience (Number of Items Produced)	Inspection Outcome
x_1	5	10	Unacceptable
x_2	5	10	Unacceptable
x_3	12	5	Unacceptable
x_4	12	12	Acceptable
x_5	12	12	Unacceptable
x_6	20	10	Acceptable
x_7	24	12	Acceptable
x_8	6	8	Unacceptable
x_9	18	10	Acceptable
x_{10}	10	10	Unacceptable

shown in Table 4.3 were collected based on inspecting ten items. For each item, the table shows the experience level of the person who made the item (measured as the number of years at the job and number of items the person has produced), and the outcome of the inspection. Representing each item by its pair of the number of months and number of items, Figure 4.14 can be developed using rough sets. The figure shows x_1, x_2, x_3, x_8 and x_{10} as items that are unacceptable, x_4 and x_5 as borderline items of the rough set that are either acceptable or unacceptable, and x_6, x_7, and x_9 as items that are acceptable. Based on this figure, the following rough set of unacceptable items (A) can be defined:

$$\underline{R}(A) = \{x_1, x_2, x_3, x_8, x_{10}\} \quad (4.97)$$
$$\overline{R}(A) = \{x_1, x_2, x_3, x_4, x_5, x_8, x_{10}\}$$

Similarly, the following rough set of acceptable items (A) can be defined:

$$\underline{R}(\overline{A}) = \{x_6, x_7, x_9\} \quad (4.98)$$
$$\overline{R}(\overline{A}) = \{x_4, x_5, x_6, x_7, x_9\}$$

These rough sets are shown in Figure 4.14.

4.2.4.4 Rough functions

The concept of rough functions was introduced by Pawlak (1999) to present a coarse approximation of unknown functions. Rough functions can be an effective means to meet the needs of engineers and scientists for our purposes in this chapter to express expert opinions. In developing relationships among variables underlying an engineering system, these rough functions can be used to articulate and express the opinion of experts in cases, such as for economic forecasts, power consumption forecasting, and assessing extreme loads on a structure.

Chapter four: Expressing and modeling expert opinions   161

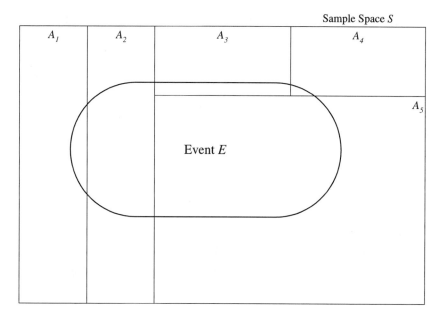

Figure 4.14 Rough sets for presenting product quality.

This section provides a format for presenting these subjective relationships. Although the mathematics needed to manipulate these functions are not provided, they can be obtained and developed based on the work of Pawlak (1999). In engineering and science, we are interested in cause-effect relationships expressed as

$$Y = f(x) \qquad (4.99)$$

where Y is the value of the criterion variable, or the dependent variable; x is the predictor variable, or the independent variable; and f is the functional relationship. Using the concept of lower and upper approximations of f, a rough function can be expressed using the following pair:

$$Y = (\underline{R}(f), \overline{R}(f)) \qquad (4.100)$$

$\underline{R}(f)$ and $\overline{R}(f)$ are lower and upper approximations of f. Rough functions can be used to extrapolate an empirical function to regions beyond data availability for developing forecasting models. Functional operations such as derivatives, integrals, roots, maximum, and minimum can be defined on the two approximations. Also, they can be performed using numerical techniques performed on values at the lower and upper approximations. Ayyub and McCuen (1996) describe commonly used numerical methods with practical examples.

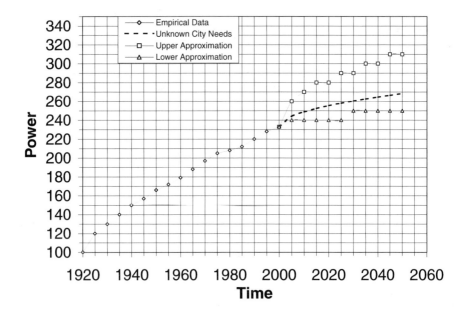

Figure 4.15 A rough function for forecasting power needs.

Example 4.14 Forecasting power needs using a rough function

The power needs of a city can be forecasted to help city planners in making zoning and power-plant construction decisions as discussed in Example 4.12. Figure 4.15 shows an empirical power consumption trend over time and a subjectively developed forecast of the city's needs using lower and upper approximations of the needs (f). The forecasted segment of the curve is provided for demonstration purposes. The rough function was developed by establishing convenient gridlines, every 10 units of power and every five years as shown Figure 4.15. Then, lower and upper approximation of the city's unknown needs for power by identifying the coordinate points that would include the unknown function.

4.3 Monotone measures

Monotone measures are often referred to in the literature as *fuzzy measures* (Wang and Klir, 1992; Klir and Wierman, 1999), although fuzzy sets are not necessarily involved in the definition of these measures. The concept of monotone measures was conceived by Sugeno (1974). Monotone measures can also be developed for fuzzy sets, and this case should be called fuzzy measures.

Monotone measures are needed to model inconsistency and incompleteness ignorance types as shown in Figure 1.9. They can be used to model uncertainty in the membership of an element of a universal set X to a family of sets that are identified by an analyst for the convenience of capturing and expressing opinions. The membership uncertainty in this case is different from the membership functions of fuzzy sets. While fuzzy sets have membership functions to represent uncertainty in belonging to a set of fuzzy boundaries or a vague concept, monotone measures provide an assessment of the membership likelihood of an element of X to each set in the family of sets identified by the analyst. The family of sets are crisp sets, whereas the element is imprecise in its boundaries and hence uncertain in its belonging to the sets of the family of sets. In fuzzy sets, the elements of X are crisp and precise, but a set of interest is vague or fuzzy, hence the uncertainty. An example of a situation where monotone measures can be used is in filing a historical failure case of a system to predetermined, precisely defined, classes of failure causes. The failure case at hand might have a complex cause that would make classifying it to only one failure-cause class unrealistic and would make its classification possible only in the form of a monotone measure, i.e., likelihood of belonging, to each of the prescribed failure classes.

4.3.1 Definition of monotone measures

A *monotone measure* can be defined based on a nonempty family A of subsets for a given universal set X, with an appropriate algebraic structure as a mapping from A to $[0,\infty]$ as follows:

$$f: A \to [0,\infty] \qquad (4.101)$$

The monotone measure must vanish at the empty set as follows:

$$f(\emptyset) = 0 \qquad (4.102)$$

For all A_1 and $A_2 \in A$, and $A_1 \subseteq A_2$, then $f(A_1) \leq f(A_2)$, i.e., monotonic sets. For any increasing sequence $A_1 \subseteq A_2 \subseteq \ldots$ of sets in A, the function f is continuous from below, and for any decreasing sequence $A_1 \supseteq A_2 \supseteq \ldots$ of sets in A, the function f is continuous from above. Functions that satisfy the above four requirements are called *semicontinuous* from below or above, respectively. For any pair A_1 and $A_2 \in A$ such that $A_1 \cap A_2 = \emptyset$, a monotone measure f can have one of the following properties as provided by Equations 4.103 – 4.105:

$$f(A_1 \cup A_2) > f(A_1) + f(A_2) \qquad (4.103)$$

Equation 4.103 can be viewed as expressing a *cooperative action* or *synergy* between A_1 and A_2.

$$f(A_1 \cup A_2) = f(A_1) + f(A_2) \qquad (4.104)$$

Equation 4.104 can be viewed as expressing a *noninteractive* nature of A_1 and A_2.

$$f(A_1 \cup A_2) < f(A_1) + f(A_2) \qquad (4.105)$$

Equation 4.105 can be viewed as expressing an inhibitory effect or incompatibility between A_1 and A_2 (Wang and Klir, 1992; Klir and Wierman, 1999). Probability theory is based on a classical measure theory as provided by Equation 4.104. Therefore, the theory of monotone measures provides a broader framework than probability theory for modeling uncertainty. In fact, monotone measures are general enough to encompass probability theory, possibility theory, and theory of evidence.

4.3.2 Classifying monotone measures

A two-way classification of monotone measures can be developed based on the work of Wang and Klir (1992) and Klir and Wierman (1999), using the following two considerations: (1) the type of sets (crisp, fuzzy and/or rough), and (2) the properties provided by Equations 4.103 – 4.105. The Cartesian combination of these two considerations would produce many types of monotone measures that are not all fully developed in the literature. The reasonably developed types of monotone measures are listed below.

1. **Classical probability theory.** Classical probability theory is based on crisp sets and additive measures, i.e., Equation 4.104.
2. **Probability theory based on fuzzy events** (fuzzy sets and additive measures, i.e., Equation 4.104). In this case classical probability concepts are extended to deal with fuzzy events as demonstrated by Ayyub and Lai (1994).
3. **Dempster-Shafer theory of evidence and its monotone measures of belief and plausibility** (crisp sets and nonadditive measures, i.e., Equations 4.103 and 4.105). This is the classic case of the Dempster-Shafer theory of evidence (Shafer, 1976; Dempster, 1976a, 1976b). The theory of evidence defines and uses two monotone measures called belief and plausibility measures that form a dual.
4. **Fuzzified Dempster-Shafer theory of evidence** (fuzzy sets and nonadditive measures, i.e., Equations 4.103 and 4.105). This is an extension of the classic case of the Dempster-Shafer theory of evidence by considering fuzzy sets.
5. **Possibility theory and its monotone measures of necessity and possibility** (crisp sets and nonadditive measures, i.e., Equations 4.103 and 4.105). This possibility theory case is a special case of the above Dempster-Shafer theory of evidence and its monotone measures of belief and plausibility by requiring underlying events to be nested, i.e., $A_1 \subset A_2 \subset \ldots \subset X$.

Chapter four: Expressing and modeling expert opinions

6. **Possibility theory based on fuzzy events** (fuzzy sets and nonadditive measures, i.e., Equations 4.103 and 4.105). This possibility theory is an extension of the classic case of the possibility theory of evidence by considering fuzzy sets.
7. **Other cases.** A large number of cases can be developed based on the nonadditive measures of Equations 4.103 and 4.105, such as imprecise probabilities, and based on rough sets. There are new theories that are emerging in these areas but are not mature enough for the purposes of this book.

In this chapter, the theory of evidence, probability theory, and possibility theory are introduced for their potential use in capturing and expressing the opinions of experts.

4.3.3 Evidence theory

The theory of evidence, also called the Dempster-Shafer theory, was developed by Shafer (1976), and Dempster (1976a, 1976b). The underlying monotone measures for this theory are the belief measure and plausibility measure.

4.3.3.1 Belief measure

The belief measure (Bel) should be defined on a universal set X as a function that maps the power set of X to the range $[0,1]$ as given by

$$Bel: P_X \rightarrow [0,1] \quad (4.106)$$

where P_X is the set of all subsets of X and is called the *power set* of X. The power set has $2^{|X|}$ subsets in it. The belief function has to meet the following three conditions:

$$Bel(\emptyset) = 0 \quad (4.107)$$

$$Bel(X) = 1 \quad (4.108)$$

$$Bel(A_1 \cup A_2 \cup \cdots \cup A_N) \geq \sum_{j=1}^{N} Bel(A_j) - \sum_{k=1}^{N}\left[\sum_{j=k+1}^{N} Bel(A_j \cap A_k)\right] + \quad (4.109)$$

$$\sum_{k=1}^{N}\left\{\sum_{j=k+1}^{N}\left[\sum_{l=j+1}^{N} Bel(A_j \cap A_k \cap A_l)\right]\right\} - \cdots + (-1)^{N+1} Bel(A_1 \cap A_2 \cap \cdots \cap A_N)$$

where $\emptyset$ is the null set, and $A_1, A_2, \ldots, A_N$ is any possible family of subsets of X. The inequality provided by Equation 4.109 shows that the belief measure has the property of being super-additive.

4.3.3.2 Plausibility measure

The belief measure (*Bel*) has a dual measure called the plausibility measure (*Pl*) as defined by the following equation:

$$Pl(A) = 1 - Bel(\overline{A}) \tag{4.110}$$

where A is a subset that belongs to the power set P_X. The plausibility measure must satisfy the following conditions:

$$Pl(\emptyset) = 0 \tag{4.111}$$

$$Pl(X) = 1 \tag{4.112}$$

$$Pl(A_1 \cap A_2 \cap \cdots \cap A_N) \leq \sum_{j=1}^{N} Pl(A_j) - \sum_{k=1}^{N}\left[\sum_{j=k}^{N} Pl(A_j \cup A_k)\right] + \cdots \tag{4.113}$$

$$+(-1)^{N+1} Pl(A_1 \cup A_2 \cup \cdots \cup A_N)$$

where $A_1, A_2, \ldots, A_N$ is any family of subsets of X. The pair, *Bel* and *Pl*, forms a duality. The inequality provided by Equation 4.113 shows that the plausibility measure has the property of being sub-additive. It can be shown that the belief and plausibility functions satisfy the following condition:

$$Pl(A) \geq Bel(A) \tag{4.114}$$

for each A in the power set.

4.3.3.3 Basic assignment

In engineering and science, a body of evidence represented by a family of sets $(A_1, A_2, \ldots, A_N)$ can be characterized by a basic assignment constructed for convenience and for facilitating the expression and modeling of, for example, expert opinions. A basic assignment can be related to the belief and plausibility measures; however, its creation is commonly easier and more relevant to problems than directly developing the belief and plausibility measures. A basic assignment provides an assessment of the membership likelihood of an element x of X to each set in a family of sets identified by the analyst. The family of sets are crisp sets, whereas the element of interest x is imprecise in its boundaries and hence uncertain in its belonging to the sets in the family of sets.

A basic assignment can be conveniently characterized by the following function that maps the power set to the range [0,1]:

$$m: P_X \to [0,1] \tag{4.115}$$

Chapter four: Expressing and modeling expert opinions 167

A basic assignment must satisfy the following two conditions:

$$m(\emptyset) = 0 \qquad (4.116)$$

$$\sum_{\text{all } A \in P_X} m(A) = 1 \qquad (4.117)$$

If $m(A_i) > 0$ for any i, A_i is called also a *focal element*. The belief measure and plausibility measure can be computed based on a particular basic assignment m, for any set $A_i \in P_X$ as follows:

$$Bel(A_i) = \sum_{\text{all } A_j \subseteq A_i} m(A_j) \qquad (4.118)$$

$$Pl(A_i) = \sum_{\text{all } A_j \cap A_i \neq \emptyset} m(A_j) \qquad (4.119)$$

The summation in Equation 4.118 should be performed over all sets A_j that are contained or equal to A_i, whereas the summation in Equation 4.119 should be performed over all sets A_j that belong to or intersect with the set A_i. The three functions, *Bel*, *Pl*, and *m*, can be viewed as alternative representations of the same information or evidence regarding the element x. These functions express the likelihood that x belongs to each A_i as a belief measure (strongest), plausibility measure (weakest), and a basic assignment (collected evidence). Once one of the three functions is defined, the other two functions can be uniquely computed. For example, the basic assignment m for $A_i \in P_X$ can be computed based on the belief function as follows:

$$m(A_i) = \sum_{\text{all } A_j \subseteq A_i} (-1)^{|A_i - A_j|} Bel(A_j) \qquad (4.120)$$

where $|A_i - A_j|$ is the cardinality of the difference between the two sets. Equation 4.110 can be used to compute the *Bel* from the *Pl* for cases where *Pl* values are given; then Equation 4.120 can be used to compute m.

Basic assignments (m_1 and m_2) produced by two experts on the same element and a family of sets of interest can be combined using Dempster's rule of combination to obtain a combined pinion ($m_{1,2}$) as follows:

$$m_{1,2}(A_i) = \frac{\sum_{\text{all } A_j \cap A_k = A_i} m_1(A_j) m_2(A_k)}{1 - \sum_{\text{all } A_j \cap A_k = \emptyset} m_1(A_j) m_2(A_k)} \qquad (4.121)$$

where A_i must be a nonempty set, and $m_{1,2}(\emptyset) = 0$. The term $1 - \sum_{\text{all } A_j \cap A_k = \emptyset} m_1(A_j) m_2(A_k)$ of Equation 4.121 is a normalization factor. Equation 4.121 provides an example rule to combine expert opinions that does not account for the reliability of the source and other relevant considerations.

Example 4.15 Causes of a bridge failure during construction

Bridges can collapse during construction due to many causes (Eldukair and Ayyub, 1991). Consider three common causes: design error (D), construction error (C), and human error (H). A database of bridges that failed during construction can be developed. For each bridge failure case, the case needs to be classified in terms of its causes and entered in the database. The sets D, C, and H belong to the universal set X of failure causes. Two experts were asked to review a bridge-failure case and subjectively provide basic assignments for this case in terms of the sets D, C, and H. The experts provided the assignments in Table 4.4 for D, C, H, D∪C, D∪H, and C∪H. The assignment for D∪C∪H was computed based on Equation 4.117 to obtain a total of one for the assignments provided by each expert. The *Bel* values for each expert in Table 4.4 were computed using Equation 4.118.

In order to combine the opinions of the experts according to Equation 4.121, the normalizing factor needs to be computed as follows:

$$1 - \sum_{\text{all } A_j \cap A_k = \emptyset} m_1(A_j) m_2(A_k) =$$

$$1 - \begin{bmatrix} m_1(D)m_2(C) + m_1(D)m_2(H) + m_1(D)m_2(C \cup H) + \\ m_1(C)m_2(D) + m_1(C)m_2(H) + m_1(C)m_2(D \cup H) + \\ m_1(H)m_2(D) + m_1(H)m_2(C) + m_1(H)m_2(C \cup D) + \\ m_1(D \cup C)m_2(H) + m_1(D \cup H)m_2(C) + m_1(C \cup H)m_2(D) \end{bmatrix} \quad (4.122)$$

substituting the values of m produces the following normalizing factor:

$$1 - \sum_{\text{all } A_j \cap A_k = \emptyset} m_1(A_j) m_2(A_k) =$$

$$1 - \begin{bmatrix} 0.1(0.1) + 0.1(0.15) + 0.1(0.1) + \\ (0.05)(0.05) + (0.05)(0.15) + (0.05)(0.1) + \\ (0.1)(0.05) + (0.1)(0.1) + (0.1)(0.25) + \\ (0.2)(0.15) + (0.1)(0.1) + (0.05)(0.05) \end{bmatrix} = 0.8675 \quad (4.123)$$

Chapter four: Expressing and modeling expert opinions

The combined opinions can then be computed using Equation 4.121 as follows:

$$m_{1,2}(D) = \frac{\begin{bmatrix} m_1(D)m_2(D) + m_1(D)m_2(D \cup C) + m_1(D)m_2(D \cup H) + \\ m_1(D)m_2(D \cup C \cup H) + m_1(D \cup C)m_2(D) + m_1(D \cup C)m_2(D \cup H) + \\ m_1(D \cup H)m_2(D) + m_1(D \cup H)m_2(D \cup C) + m_1(D \cup C \cup H)m_2(D) \end{bmatrix}}{0.8675}$$

(4.124)

or

$$m_{1,2}(D) = \frac{\begin{bmatrix} 0.1(0.5) + 0.1(0.25) + 0.1(0.1) + \\ 0.1(0.25) + 0.2(0.05) + 0.2(0.1) + \\ 0.1(0.05) + 0.1(0.25) + 0.4(0.05) \end{bmatrix}}{0.8675} = 0.167147 \quad (4.125)$$

$$m_{1,2}(C) = \frac{\begin{bmatrix} 0.05(0.1) + 0.05(0.25) + 0.05(0.1) + \\ 0.05(0.25) + 0.2(0.1) + 0.2(0.1) + \\ 0.05(0.1) + 0.05(0.25) + 0.4(0.1) \end{bmatrix}}{0.8675} = 0.152738 \quad (4.126)$$

$$m_{1,2}(H) = \frac{\begin{bmatrix} 0.1(0.15) + 0.1(0.1) + 0.1(0.1) + \\ 0.1(0.25) + 0.1(0.15) + 0.1(0.1) + \\ 0.05(0.15) + 0.05(0.1) + 0.4(0.15) \end{bmatrix}}{0.8675} = 0.181556 \quad (4.127)$$

$$m_{1,2}(D \cup C) = \frac{[0.2(0.25) + 0.2(0.25) + 0.4(0.25)]}{0.8675} = 0.230548 \quad (4.128)$$

$$m_{1,2}(D \cup H) = \frac{[0.1(0.1) + 0.1(0.25) + 0.4(0.1)]}{0.8675} = 0.086455 \quad (4.129)$$

$$m_{1,2}(C \cup H) = \frac{[0.05(0.1) + 0.05(0.25) + 0.4(0.1)]}{0.8675} = 0.066282 \quad (4.130)$$

$$m_{1,2}(C \cup D \cup H) = \frac{[0.4(0.25)]}{0.8675} = 0.115274 \quad (4.131)$$

Table 4.4 Belief Computations for Classifying Bridge Failures

Subset* (i.e., failure cause)	Expert 1		Expert 2		Combined Judgment	
	m_1	Bel_1	m_2	Bel_2	$m_{1,2}$	$Bel_{1,2}$
Design error (D)	0.10	0.10	0.05	0.05	0.167147	0.167147
Construction error (C)	0.05	0.05	0.10	0.10	0.152738	0.152738
Human error (H)	0.10	0.10	0.15	0.15	0.181556	0.181556
$D \cup C$	0.20	0.35	0.25	0.40	0.230548	0.550433
$D \cup H$	0.10	0.30	0.10	0.30	0.086455	0.435158
$C \cup H$	0.05	0.20	0.10	0.35	0.066282	0.400576
$D \cup C \cup H$	0.40	1.	0.25	1.	0.115274	1.

*The subsets could also be written as {D}, {C}, {H}, {D,C}, {D,H}, {C,H}, and {D,C,H}, respectively.

Table 4.5 Plausibility Computations for Classifying Bridge Failures

Subset* (i.e., failure cause)	Expert 1		Expert 2		Combined Judgment	
	m_1	Pl_1	m_2	Pl_2	$m_{1,2}$	$Pl_{1,2}$
Design error (D)	0.10	0.80	0.05	0.65	0.167147	0.599424
Construction error (C)	0.05	0.70	0.10	0.70	0.152738	0.564842
Human error (H)	0.10	0.65	0.15	0.60	0.181556	0.449567
$D \cup C$	0.20	0.90	0.25	0.85	0.230548	0.818444
$D \cup H$	0.10	0.95	0.10	0.90	0.086455	0.847262
$C \cup H$	0.05	0.90	0.10	0.95	0.066282	0.832853
$D \cup C \cup H$	0.40	1.	0.25	1.	0.115274	1.

*The subsets could also be written as {D}, {C}, {H}, {D,C}, {D,H}, {C,H}, and {D,C,H}, respectively.

The values provided by Equations 4.125 – 4.131 must add up to one. The $Bel_{1,2}$ values in Table 4.4 were computed using Equation 4.118. The plausibility computations for classifying bridge failures are shown in Table 4.5 that were based on Equation 4.119.

4.3.4 Probability theory

4.3.4.1 Relationship between evidence theory and probability theory

Probability theory can be treated as a special case of the theory of evidence. For cases in which all focal elements for a given basic assignment, m (i.e., body of evidence) are singletons, the associated belief measure and plausibility measure collapse into a single measure, a classical probability measure. The term *singleton* means that each subset A_i of the family A of subsets, presenting an evidence body, contains only one element. The resulting probability measure is additive in this case; i.e., it follows Equation 4.104. The following differences between evidence theory and probability theory can be noted based on this reduction of evidence theory to probability theory:

- A basic assignment in evidence theory can be used to compute the belief and plausibility measures that map the power set of X to the range [0,1].
- A probability assignment, such as a probability mass function, in probability theory is a mapping function from the universal set X to the range [0,1].

Dempster (1967a and 1967b) examined the use of evidence theory to construct a probability distribution for singletons based on a basic assignment for some subsets of a universal set. The solution can be expressed in the form of minimum and maximum probabilities for the singletons for cases where the evidence body (i.e., the basic assignment) not contradictory. This construct is not covered in this book.

4.3.4.2 Classical definitions of probability

The concept of probability has its origin in games of chance. In these games, probabilities are determined based on many repetitions of an experiment and counting the number of outcomes of an event of interest. Then, the probability of the outcome of interest can be measured by dividing the number of occurrences of an event of interest by the total number of repetitions. Quite often, probability is specified as a percentage; for example, when the weather bureau indicates that there is a 30 percent chance of rain, experience indicates that under similar meteorological conditions it has rained 3 out of 10 times. In this example, the probability was estimated empirically using the concept of relative frequency expressed as

$$P(X = x) = \frac{n}{N} \qquad (4.132)$$

in which n = number of observations on the random variable X that results in an outcome of interest x, and N = total number of observations of x. The probability value associated with an event x in this equation was defined as the *relative frequency* of its occurrence. Also, probability can be defined as a *subjective probability* (or called *judgmental probability*) of the occurrence of the event. The type of definition depends on the nature of the underlying event. For example, in an experiment that can be repeated N times with n occurrences of the underlying event, the relative frequency of occurrence can be considered as an estimate of the probability of occurrence. In this case, the probability of occurrence is n/N. However, many engineering problems do not involve large numbers of repetitions, and still we are interested in estimating the probability of occurrence of some event. For example, during the service life of an engineering product, the product either fails or does not fail in performing a set of performance criteria. The events of unsatisfactory performance and satisfactory performance are mutually exclusive and collectively exhaustive of the sample space (that is the space of all possible outcomes). The probability of unsatisfactory performance (or satisfactory performance) can be considered as a subjective probability. Another example is the failure probability of a dam because of an extreme flooding condition. Estimates of such probabilities can be achieved by modeling the underlying system, its uncertainties, and performances. The resulting subjective probability is expected to reflect the status of our knowledge about the system regarding occurrence of the events of interest. Therefore, subjective probabilities can be associated with degrees of belief and can form a basis for Bayesian methods (Ayyub and McCuen, 1997). It is important to keep in mind both definitions, so results are not interpreted beyond the range of their validity.

An axiomatic definition of probability is commonly provided in the literature such as Ayyub and McCuen (1997). For an event A, the notation $P(A)$ means the probability of occurrence of the event A. The probability P should satisfy the following axioms:

1. $0 \leq P(A) \leq 1$, for any A that belongs to the *set* of all possible outcomes (called sample space S) for the system.
2. The probability of having S, $P(S) = 1$.
3. The occurrence probability of the union of mutually exclusive events is the sum of their individual occurrence probabilities.

The first axiom states that the probability of any event is inclusively between 0 and 1. Therefore, negative probabilities, or probabilities larger than one, are not allowed. The second axiom comes from the definition of the sample space. Since the sample space is the set of all possible outcomes, one or more of these outcomes must occur resulting in the occurrence of S. If the probability of the sample space does not equal 1, the sample space was incorrectly defined. The third axiom sets a basis for the mathematics of probability. These axioms as a single entity can be viewed as a definition of probability; i.e., any numerical structure that adheres to these axioms will provide a probability structure. Therefore, the relative frequency and subjective probability meet this definition of probability.

Relative frequency and Subjective probability are tools that help engineers and planners to deal with and model uncertainty and should be used appropriately as engineering systems and models demand. In the case of relative frequency, increasing the number of repetitions according to Equation 4.132 would produce an improved estimate with a diminishing return on invested computational and experimental resources until a limiting (i.e. long-run or long-term) frequency value is obtained. This limiting value can be viewed as the *true probability* although the absolute connotation in this terminology might not be realistic and cannot be validated. Philosophically, a true probability might not exist especially when dealing with subjective probabilities. This, however, does not diminish the value of probabilistic analysis and methods since they provide a consistent, systematic, rigorous, and robust framework for dealing with uncertainty and decision making.

4.3.4.3 Linguistic probabilities

Probability as described in the previous section provides a measure of the likelihood of occurrence of an event. It is a numerical expression of uncertainty; however, it is common for subjects (such as experts) to express uncertainty verbally using linguistic terms, such as likely, probable, and improbable. Although, these linguistic terms are somewhat fuzzy, they are meaningful. Lichtenstein and Newman (1967) developed a table that translates commonly used linguistic terms into probability values using responses from subjects; a summary of which is shown in Table 4.6 (Beacher, 1998). The responses of the subjects show encouraging consistency in defining each term; however, the range of responses is large. Moreover, mirror-image pairs sometimes produce asymmetric results. The term "rather unlikely" is repeated in the table as it was used twice in the questionnaire, at the start and end, to check consistency. It can be concluded from this table that verbal descriptions of uncertainty can be useful as an initial assessment, but other analytical techniques should be used to assess uncertainty; for example, the linguistic terms in Table 4.6 can be modeled using fuzzy sets (Haldar et al., 1997; Ayyub et al., 1997; Ayyub and Gupta, 1997; Ayyub, 1998).

4.3.4.4 Failure rates

A failure rate can be defined as the probability of failure per unit time or a unit of operation, such as cycle, revolution, rotation, or start-up. For example, a constant failure rate for an electronic device of 0.1 per year means that on the average, the device fails once per 10 years. Another example that does not involve time is an engine with a failure rate of 10^{-5} per cycle of operation (or it can be in terms of mission length). In this case, the failure rate means that on the average the engine fails once per 100,000 cycles. Due to manufacturing, assembly, and aging effects, failure rates can generally vary with time (or other units of operation), therefore, requiring sometimes a statement of limitation on their applicability. Failure rates can be used in probabilistic analysis. There are analytical methods to convert failure rates into probabilities of some events of interest.

4.3.4.5 Central tendency measures

A very important descriptor of data is central tendency measures. The central tendency can be measured using, for example, the mean (or average) value or the median value.

4.3.4.5.1 Mean (or average) value.
The average value is the most commonly used central tendency descriptor. The definition of the mean (or average) value herein is based on a sample of size n. The sample consists of n values of a random variable X. For n observations, if all observations are given equal weight, then the average value is given by

Table 4.6 Linguistic Probabilities and Translations

Rank	Phrase	No. of Responses	Mean	Median	Standard Deviation	Range
1	Highly probable	187	0.89	0.90	0.04	0.60-0.99
2	Very likely	185	0.87	0.90	0.06	0.60-0.99
3	Very probable	187	0.87	0.89	0.07	0.60-0.99
4	Quite likely	188	0.79	0.80	0.10	0.30-0.99
5	Usually	187	0.77	0.75	0.13	0.15-0.99
6	Good chance	188	0.74	0.75	0.12	0.25-0.95
7	Predictable	146	0.74	0.75	0.20	0.25-0.95
8	Likely	188	0.72	0.75	0.11	0.25-0.99
9	Probable	188	0.71	0.75	0.17	0.01-0.99
10	Rather likely	188	0.69	0.70	0.09	0.15-0.99
11	Pretty good chance	188	0.67	0.70	0.12	0.25-0.95
12	Fairly likely	188	0.66	0.70	0.12	0.15-0.95
13	Somewhat likely	187	0.59	0.60	0.18	0.20-0.92
14	Better than even	187	0.58	0.60	0.06	0.45-0.89
15	Rather	124	0.58	0.60	0.11	0.10-0.80
16	Slightly more than half the time	188	0.55	0.55	0.06	0.45-0.80
17	Slight odds in favor	187	0.55	0.55	0.08	0.05-0.75
18	Fair chance	188	0.51	0.50	0.13	0.20-0.85
19	Tossup	188	0.50	0.50	0.00	0.45-0.52
20	Fighting chance	186	0.47	0.50	0.17	0.05-0.90
21	Slightly less than half the time	188	0.45	0.45	0.04	0.05-0.50
22	Slight odds against	185	0.45	0.45	0.11	0.10-0.99
23	Not quite even	180	0.44	0.45	0.07	0.05-0.60
24	Inconclusive	153	0.43	0.50	0.14	0.01-0.75

Table 4.6 continued

25	Uncertain	173	0.40	0.50	0.14	0.08-0.90
26	Possible	178	0.37	0.49	0.23	0.01-0.99
27	Somewhat unlikely	186	0.31	0.33	0.12	0.03-0.80
28	Fairly unlikely	187	0.25	0.25	0.11	0.02-0.75
29	Rather unlikely	187	0.24	0.25	0.12	0.01-0.75
30	Rather unlikely	187	0.21	0.20	0.10	0.01-0.75
31	Not very probable	187	0.20	0.20	0.12	0.01-0.60
32	Unlikely	188	0.18	0.16	0.10	0.01-0.45
33	Not much chance	186	0.16	0.15	0.09	0.01-0.45
34	Seldom	188	0.16	0.15	0.08	0.01-0.47
35	Barely possible	180	0.13	0.05	0.17	0.01-0.60
36	Faintly possible	184	0.13	0.05	0.16	0.01-0.50
37	Improbable	187	0.12	0.10	0.09	0.01-0.40
38	Quite unlikely	187	0.11	0.10	0.08	0.01-0.50
39	Very unlikely	186	0.09	0.10	0.07	0.01-0.50
40	Rare	187	0.07	0.05	0.07	0.01-0.30
41	Highly improbable	181	0.06	0.05	0.05	0.01-0.30

Lichtenstein and Newman, 1967

$$\overline{X} = \frac{1}{n}\sum_{i=1}^{n} x_i \qquad (4.133)$$

where x_i = a sample point, and $i = 1, 2, \ldots, n$; and

$$\sum_{i=1}^{n} x_i = x_1 + x_2 + x_3 + \ldots + x_n \qquad (4.134)$$

Since the average value ($\overline{X}$) is based on a sample, it has statistical error for two reasons: (1) it is sample dependent (i.e., a different sample might produce a different average), and (2) it is sample-size dependent (i.e., as the sample size is increased, the error is expected to reduce). The mean value has another mathematical definition that is based on probability distributions according to probability theory, which is not described herein.

4.3.4.5.2 *Average time between failures.* The average time between failures can be computed as the average ($\overline{X}$), where x_i = a sample point indicating the age at failure of a failed component, and $i = 1, 2, \ldots, n$. The failed components are assumed to be replaced by new identical ones or repaired to a state "as good as new." The average time between failures is related to the failure rate as its reciprocal. For example, a component with a failure rate of 0.1 per year has an average time between failures of 1/0.1 = 10 years. Similar to failure rates, the average time between failures can be constant or time dependent.

4.3.4.5.3 *Median value.* The median value x_m is another measure of central tendency. It is defined as the point that divides the data into two equal parts; i.e., 50% of the data are above x_m and 50% are below x_m. The median value can be determined by ranking the n values in the sample in decreasing order, 1 to n. If n is an odd number, then the median is the value with a rank of $(n+1)/2$. If n is an even number, then the median equals the average of the two middle values, i.e., those with ranks $n/2$ and $(n/2)+1$.

The advantage of using the median value as a measure of central tendency over the average value is its insensitivity to extreme values, such as outliers. Consequently, this measure of central tendency is commonly used in combining expert judgments in an expert-opinion elicitation process.

4.3.4.6 Dispersion (or variability)

Although the central tendency measures convey certain information about the underlying sample, they do not completely characterize the sample. Two random variables can have the same mean value, but different levels of data scatter around the computed mean. Thus, measures of central tendency cannot fully characterize the data. Other characteristics are also important

Chapter four: Expressing and modeling expert opinions

and necessary. The dispersion measures describe the level of scatter in the data about the central tendency location.

The most commonly used measure of dispersion is the variance and other quantities that are derived from it, such as the standard deviation and coefficient of variation. For n observations in a sample that are given equal weight, the variance (S^2) is given by

$$S^2 = \frac{1}{n-1}\sum_{i=1}^{n}(x_i - \overline{X})^2 \qquad (4.135)$$

The units of the variance are the square of the units of the variable x; for example, if the variable is measured in pounds per square inch (psi), the variance has units of (psi)². Computationally, the variance of a sample can be determined using the following alternative equation:

$$S^2 = \frac{1}{n-1}\left[\sum_{i=1}^{n}x_i^2 - \frac{1}{n}\left(\sum_{i=1}^{n}x_i\right)^2\right] \qquad (4.136)$$

By definition, the standard deviation (S) is the square root of the variance as follows:

$$S = \sqrt{\frac{1}{n-1}\left[\sum_{i=1}^{n}x_i^2 - \frac{1}{n}\left(\sum_{i=1}^{n}x_i\right)^2\right]} \qquad (4.137)$$

It has the same units as both the underlying variable and the central tendency measures. Therefore, it is a useful descriptor of the dispersion or spread of a sample of data. The coefficient of variation (COV) is a normalized quantity based on the standard deviation and the mean and is different from the covariance. The covariance measures any association between two random variables. Therefore, the COV is dimensionless and is defined as

$$COV = \frac{S}{\overline{X}} \qquad (4.138)$$

It is also used as an expression of the standard deviation in the form of a percent of the average value. For example, consider $\overline{X}$ and S to be 50 and 20, respectively; therefore, $COV(X) = 0.4$ or 40%. In this case, the standard deviation is 40% of the average value.

4.3.4.7 Percentiles

A p-percentile value (x_p) for a random variable based on a sample is the value of the parameter such that p% of the data is less or equal to x_p. On the basis of this definition, the median value is considered to be the 50-percentile value.

Aggregating the opinions of experts sometimes requires the computation of the 25, 50, and 75 percentile values. The computation of these values depends on the number of experts providing opinions. Table 4.7 provides a summary of the equations needed for 4 to 20 experts. In the table, X_i means the opinion of an expert with the ith smallest value; i.e., $X_1 \geq X_2 \geq X_3 \geq \ldots \geq X_n$, where n = number of experts. In the table, the arithmetic average was used to compute the percentiles. In some cases, where the values of X_i differ by power order of magnitude, the geometric average can be used. Expert opinions should not be aggregated in this manner all the times; other aggregation methods, as provided in Chapter 5, might be more appropriate and should be considered.

4.3.4.8 Statistical uncertainty

Values of random variables obtained from sample measurements are commonly used in making important engineering decisions. For example, samples of river water are collected to estimate the average level of a pollutant in the entire river at that location. Samples of stopping distances are used to develop a relationship between the speed of a car at the time the brakes are applied and the distance traveled before the car comes to a complete halt. The average of sample measurements of the compressive strength of concrete collected during the pouring of a large concrete slab, such as the deck of a parking garage, is used to help decide whether or not the deck has the strength specified in the design specifications. It is important to recognize the random variables involved in these cases. In each case, the individual measurements or samples are values of a random variable, and the computed mean is also the value of a random variable. For example, the transportation engineer measures the stopping distance; each measurement is a sample value of the random variable. If ten measurements are made for a car stopping from a speed of 50 mph, then the sample consists of ten values of the random variable. Thus, there are two random variables in this example: the stopping distance and the estimated mean of the stopping distance; this is also true for the water-quality-pollutant and compressive-strength examples.

The estimated mean for a random variable is considered by itself to be a random variable because different samples about the random variable can produce different estimated mean values; hence the randomness in the estimated mean. When a sample of n measurements of a random variable is collected, the n values are not necessarily identical. The sample is characterized by variation. For example, assume that five independent estimates of the compressive strength of the concrete in a parking garage deck are obtained from samples of the concrete obtained when the concrete was poured. For illustration purposes, assume that the five compressive strength measurements are 3250, 3610, 3460, 3380, and 3510 psi. These measurements produce a mean of 3442 psi and a standard deviation of 135.9 psi. Assume that another sample of five measurements of concrete strength is obtained

Chapter four: Expressing and modeling expert opinions 179

Table 4.7 Computations of Percentiles

Number of experts (n)	25 percentile Arithmetic Average	25 percentile Geometric Average	50 percentile Arithmetic Average	50 percentile Geometric Average	75 percentile Arithmetic Average	75 percentile Geometric Average
4	$(X_1+X_2)/2$	$\sqrt{X_1 X_2}$	$(X_2+X_3)/2$	$\sqrt{X_2 X_3}$	$(X_3+X_4)/2$	$\sqrt{X_3 X_4}$
5	X_2	X_2	X_3	X_3	X_4	X_4
6	X_2	X_2	$(X_3+X_4)/2$	$\sqrt{X_3 X_4}$	X_5	X_5
7	$(X_2+X_3)/2$	$\sqrt{X_2 X_3}$	X_4	X_4	$(X_5+X_6)/2$	$\sqrt{X_5 X_6}$
8	$(X_2+X_3)/2$	$\sqrt{X_2 X_3}$	$(X_4+X_5)/2$	$\sqrt{X_4 X_5}$	$(X_6+X_7)/2$	$\sqrt{X_6 X_7}$
9	$(X_2+X_3)/2$	$\sqrt{X_2 X_3}$	X_5	X_5	$(X_7+X_8)/2$	$\sqrt{X_7 X_8}$
10	$(X_2+X_3)/2$	$\sqrt{X_2 X_3}$	$(X_5+X_6)/2$	$\sqrt{X_5 X_6}$	$(X_8+X_9)/2$	$\sqrt{X_8 X_9}$
11	X_3	X_3	X_6	X_6	X_9	X_9
12	X_3	X_3	$(X_6+X_7)/2$	$\sqrt{X_6 X_7}$	X_{10}	X_{10}

Table 4.7 continued

13	$(X_3+X_4)/2$	$\sqrt{X_3 X_4}$	X_7	X_7	$(X_{10}+X_{11})/2$	$\sqrt{X_{10} X_{11}}$
14	$(X_3+X_4)/2$	$\sqrt{X_3 X_4}$	$(X_7+X_8)/2$	$\sqrt{X_7 X_8}$	$(X_{11}+X_{12})/2$	$\sqrt{X_{11} X_{12}}$
15	X_4	X_4	X_8	X_8	X_{12}	X_{12}
16	X_4	X_4	$(X_8+X_9)/2$	$\sqrt{X_8 X_9}$	X_{13}	X_{13}
17	$(X_4+X_5)/2$	$\sqrt{X_4 X_5}$	X_9	X_9	$(X_{13}+X_{14})/2$	$\sqrt{X_{13} X_{14}}$
18	$(X_4+X_5)/2$	$\sqrt{X_4 X_5}$	$(X_9+X_{10})/2$	$\sqrt{X_9 X_{10}}$	$(X_{14}+X_{15})/2$	$\sqrt{X_{14} X_{15}}$
19	X_5	X_5	X_{10}	X_{10}	X_{15}	X_{15}
20	X_5	X_5	$(X_{10}+X_{11})/2$	$\sqrt{X_{10} X_{11}}$	X_{15}	X_{15}

from the same concrete pour; however, the values are 3650, 3360, 3328, 3420, and 3260 psi. In this case, the estimated mean and standard deviation are 3404, and 149.3 psi, respectively. Therefore, the individual measurement and the mean are values of two different random variables, i.e., X and $\overline{X}$.

It would greatly simplify decision making if the sample measurements were identical, i.e., if there were no sampling variation so the standard deviation would be zero. Unfortunately, that is never the case, so decisions must be made in the presence of uncertainty. For example, assume in the parking garage example that the building code requires a mean compressive strength of 3500 psi. Since the mean of 3442 psi based on the first sample is less than the required 3500 psi, should we conclude that the garage deck does not meet the design specifications? Unfortunately, decision making is not that simple. If a third sample of five measurements is randomly collected from other locations on the garage deck, the collected values are just as likely to be: 3720, 3440, 3590, 3270, and 3610 psi. This sample of five produces a mean of 3526 psi and a standard deviation of 174.4 psi. In this case, the mean exceeds the design standard of 3500 psi. Since the sample mean is greater than the specified value of 3500 psi, can we conclude that the concrete is of adequate strength? Unfortunately, we cannot conclude with certainty that the strength is adequate any more than we could conclude with the first sample that the strength was inadequate. The fact that different samples lead to different means is an indication that we cannot conclude that the design specification is not met just because the sample mean is less than the design standard. We need to have more assurance.

The data that are collected on some variable or parameter represent sample information, but it is not complete by itself, and predictions are not made directly from the sample. The intermediate step between sampling and prediction is the identification of the underlying population. The sample is used to identify the population, and then the population is used to make predictions or decisions. This sample-to-population-to-prediction sequence is true for the univariate methods in statistics.

The need, then, is for a systematic decision process that takes into account the variation that can be expected from one sample to another. The decision process must also be able to reflect the risk of making an incorrect decision. This decision making can be made using statistical methods, for example, hypothesis testing as described by Ayyub and McCuen (1997).

4.3.4.9 Bayesian methods

Engineers commonly need to solve a problem and make decisions based on limited information about one or more of the parameters of the problem. The types of information available to them can be classified using commonly used terminology in the Bayesian literature as follows:

- *objective* or empirical information based on experimental results, or observations; and

- *subjective* information based on experience, intuition, other previous problems that are similar to the one under consideration, or the physics of the problem.

The first type of information can be dealt with using the theories of probability and statistics as described in previous chapters. In this type, probability is interpreted as the frequency of occurrence assuming sufficient repetitions of the problem, its outcomes, and parameters, as a basis of the information. The second type of information is subjective and can depend on the engineer or analyst studying the problem. In this type, uncertainty that exists needs to be dealt with using probabilities. However, the definition of probability is not the same as the first type because it is viewed herein as a subjective probability that reflects the state of knowledge of the engineer or the analyst.

4.3.4.9.1 *Bayesian probabilities.* It is common in engineering to encounter problems with both objective and subjective types of information. In these cases, it is desirable to utilize both types of information to obtain solutions or make decisions. The subjective probabilities are assumed to constitute a prior knowledge about a parameter, with gained objective information (or probabilities). Combining the two types produces posterior knowledge. The combination is performed based on Bayes' theorem as described by Ayyub and McCuen (1997). If $A_1, A_2, \ldots, A_n$ represent the prior (subjective) information, or a partition of a sample space S, and $E \subset S$ represents the objective information (or arbitrary event) as shown in Figure 4.16, the theorem of total probability states that

$$P(E) = P(A_1) P(E|A_1) + P(A_2) P(E|A_2) + \ldots + P(A_n) P(E|A_n) \quad (4.139)$$

where $P(A_i)$ the probability of the event A_i, and $E|A_i$ the occurrence of E given A_i, for $i = 1, 2, \ldots, n$. This theorem is very important in computing the probability of the event E, especially in practical cases when the probability cannot be computed directly, but the probabilities of the partitioning events and the conditional probabilities can be computed.

Bayes' theorem is based on the same conditions of partitioning and events as the theorem of total probability and is very useful in computing the posterior (or reverse) probability of the type $P(A_i|E)$, for $i = 1, 2, \ldots, n$. The posterior probability can be computed as follows:

$$P(A_i|E) = \frac{P(A_i)P(E|A_i)}{P(A_1)P(E|A_1) + P(A_2)P(E|A_2) + \ldots + P(A_n)P(E|A_n)} \quad (4.140)$$

The denominator of this equation is $P(E)$, which is based on the theorem of total probability. According to Equation 4.139, the prior knowledge, $P(A_i)$,

Chapter four: Expressing and modeling expert opinions   183

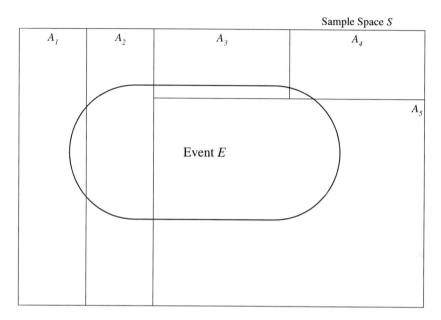

Figure 4.16 Bayes' theorem.

is updated using the objective information, $P(E)$, to obtain the posterior knowledge, $P(A_i|E)$.

Example 4.16 Defective products in manufacturing lines

A factory has three production lines. The three lines manufacture 20%, 30%, and 50% of the components produced by the factory, respectively. The quality assurance department of the factory determined that the probability of having defective products from lines 1, 2, and 3 are 0.1, 0.1, and 0.2, respectively. The following events were defined:

$$L_1 = \text{Component produced by line 1} \qquad (4.141a)$$

$$L_2 = \text{Component produced by line 2} \qquad (4.141b)$$

$$L_3 = \text{Component produced by line 3} \qquad (4.141c)$$

$$D = \text{Defective component} \qquad (4.141d)$$

Therefore, the following probabilities are given:

$$P(D|L_1) = 0.1 \qquad (4.142a)$$

$$P(D|L_2) = 0.1 \tag{4.142b}$$

$$P(D|L_3) = 0.2 \tag{4.142c}$$

Since these events are not independent, the joint probabilities can be determined as follows:

$$P(D \cap L_1) = P(D|L_1) P(L_1) = 0.1(0.2) = 0.02 \tag{4.143a}$$

$$P(D \cap L_2) = P(D|L_2) P(L_2) = 0.1(0.3) = 0.03 \tag{4.143b}$$

$$P(D \cap L_3) = P(D|L_3) P(L_3) = 0.2(0.5) = 0.1 \tag{4.143c}$$

The theorem of total probability can be used to determine the probability of a defective component as follows:

$$P(D) = P(D|L_1) P(L_1) + P(D|L_2) P(L_2) + P(D|L_3) P(L_3)$$

$$= 0.1(0.2) + 0.1(0.3) + 0.2(0.5) = 0.02 + 0.03 + 0.1$$

$$= 0.15 \tag{4.144}$$

Therefore on the average, 15% of the components produced by the factory are defective.

Because of the high contribution of Line 3 to the defective probability, a quality assurance engineer subjected the line to further analysis. The defective probability for Line 3 was assumed to be 0.2. An examination of the source of this probability revealed that it is subjective and also is uncertain. A better description of this probability can be as shown in Figure 4.17 in the form of a prior discrete distribution for the probability. The distribution is called $P_p(p)$. The mean defective component probability $\bar{p}(D)$ based on this distribution is

$$\bar{p}(D) = 0.1(0.45) + 0.2(0.43) + 0.4(0.05) + 0.6(0.04) + 0.8(0.02) + 0.9(0.01)$$
$$= 0.200 \tag{4.145}$$

Now assume that a component from Line 3 was tested and found to be defective, the subjective prior distribution of Figure 4.17 needs to be revised to reflect the new (objective) information. The revised distribution is called the posterior distribution ($P'_p(p)$), and can be computed using Equation 4.140 as follows:

$$P'_p(0.1) = \frac{0.45(0.1)}{0.2} = 0.225 \tag{4.146a}$$

Chapter four: Expressing and modeling expert opinions

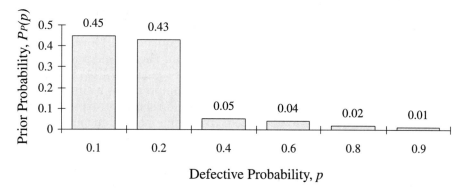

Figure 4.17 Prior probability distribution for defective probability of line 3.

Similarly, the following posterior probabilities can be computed:

$$P_p'(0.2) = \frac{0.43(0.2)}{0.2} = 0.430 \tag{4.146b}$$

$$P_p'(0.4) = \frac{0.05(0.4)}{0.2} = 0.100 \tag{4.146c}$$

$$P_p'(0.6) = \frac{0.04(0.6)}{0.2} = 0.120 \tag{4.146d}$$

$$P_p'(0.8) = \frac{0.02(0.8)}{0.2} = 0.08 \tag{4.146e}$$

$$P_p'(0.9) = \frac{0.01(0.9)}{0.2} = 0.045 \tag{4.146f}$$

The resulting probabilities in Equations 4.146 add up to 1. Also, the average probability of 0.2 can be viewed as a normalizing factor for computing these probabilities. The mean defective component probability $\bar{p}(D)$ based on the posterior distribution is

$$\bar{p}(D) = 0.1(0.225) + 0.2(0.430) + 0.4(0.100) + 0.6(0.120) \\ + 0.8(0.080) + 0.9(0.045) = 0.325 \tag{4.147}$$

The posterior mean probability (0.325) is larger than the prior mean probability (0.200). The increase is due to the failure detected by testing. Now, assume that a second component from Line 3 was tested and found to be defective; the posterior distribution of Equations 4.146 needs to be revised to reflect the new (objective) information. The revised posterior distribution builds on the posterior distribution of Equations 4.146 by treating it as a prior distribution. Performing similar computations as in Equations 4.146 and 4.147 results in the posterior distribution shown in Table 4.8 in the column "Post. 2 D." The average defective component probability $\bar{p}(D)$ is also given in the table. The last row in the table is the average nondefective component probability $(\bar{p}(ND))$ for cases where a nondefective component results from a test. This value $\bar{p}(ND)$ can be computed similar to Equation 4.145 or 4.147. For example, the $\bar{p}(ND)$ in case of a nondefective test according to the prior distribution is

$$\bar{p}(ND) = (1 - 0.1)0.225 + (1 - 0.2)0.430 + (1 - 0.4)0.100 + \\ (1 - 0.6)0.120 + (1 - 0.8)0.080 + (1 - 0.9)0.045 = 0.675 \qquad (4.148)$$

The computations for other cases are similarly performed as shown in Table 4.8. It should be noted that

$$\bar{p}(D) + \bar{p}(ND) = 1.0 \qquad (4.149)$$

Now assume that a third component from Line 3 was tested and found to be nondefective; the posterior distribution in column "Post. 2 D" of Table 4.8 needs to be revised to reflect the new (objective) information. The revised distribution is the posterior distribution $(P'_p(p))$ and can be computed using Equation 4.140 as follows:

$$P'_p(0.1) = \frac{0.0692(1 - 0.1)}{0.4883} = 0.1275 \qquad (4.150a)$$

Similarly, the following posterior probabilities can be computed:

$$P'_p(0.2) = \frac{0.2646(1 - 0.2)}{0.4883} = 0.4335 \qquad (4.150b)$$

$$P'_p(0.4) = \frac{0.1231(1 - 0.4)}{0.4883} = 0.1512 \qquad (4.150c)$$

$$P'_p(0.6) = \frac{0.2215(1 - 0.6)}{0.4883} = 0.1815 \qquad (4.150d)$$

Chapter four: Expressing and modeling expert opinions 187

Table 4.8 Prior and Posterior Distributions for Line 3

Probability, p	$P(p)$	Post. 1 D	Post. 2 D	Post. 3 ND	Post. 4 D	Post. 5 D	Post. 6 D	Post. 7 D	Post. 8 D	Post. 9 D	Post. 10 D
0.1	0.45	0.225	0.0692308	0.127599244	0.035809	0.0070718	0.0011355	0.0001638	2.22693E-05	2.912E-06	3.703E-07
0.2	0.43	0.43	0.2646154	0.433522369	0.2433245	0.0961062	0.0308633	0.0089068	0.002421135	0.0006332	0.0001611
0.4	0.05	0.1	0.1230769	0.151228733	0.1697613	0.1341016	0.08613	0.0497125	0.027026626	0.0141359	0.0071914
0.6	0.04	0.12	0.2215385	0.18147448	0.3055703	0.3620744	0.3488266	0.3020033	0.246280127	0.1932203	0.1474458
0.8	0.02	0.08	0.1969231	0.080655325	0.1810787	0.2860835	0.3674882	0.4242131	0.461254413	0.482506	0.4909318
0.9	0.01	0.045	0.1246154	0.025519849	0.0644562	0.1145626	0.1655564	0.2150004	0.262995429	0.3095016	0.3542696
$\bar{p}(D)$	0.2	0.325	0.5116923	0.356332703	0.506366	0.6227868	0.6930255	0.7357556	0.764764597	0.7862698	0.8029643
$\bar{p}(ND)$	0.8	0.675	0.4883077	0.643667297	0.493634	0.3772132	0.3069745	0.2642444	0.235235403	0.2137302	0.1970357

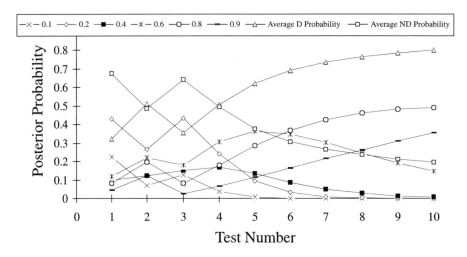

Figure 4.18 Posterior distributions for line 3.

$$P'_p(0.8) = \frac{0.1969(1-0.8)}{0.4883} = 0.0807 \qquad (4.150e)$$

$$P'_p(0.9) = \frac{0.12461(1-0.9)}{0.4883} = 0.0255 \qquad (4.150f)$$

The resulting probabilities in Equations 4.150 add up to 1. The probability $\bar{p}(ND)$ of 0.4883 was used in these calculations. The results of these calculations and the mean probability $\bar{p}(D)$ are shown in Table 4.8. It can be noted from the table that the mean defective component probability decreases as nondefective components are obtained through testing.

If the next sevenstests result in defective components, the resulting posterior distributions are shown in Table 4.8. The results are also shown in Figure 4.18. It can be observed from the figure that the average probability is approaching 1 as more and more defective tests are obtained. Also, the effect of a nondefective component on the posterior probabilities can be seen in this figure.

4.3.4.10 Interval probabilities

The term *interval probabilities* has more than one meaning as reported by Moore (1979), Dempster (1976a, 1976b), Cui and Blockley (1990), and Ferson et al. (1999). Various models for dealing with interval probabilities are provided in subsequent sections, but this section summarizes the model suggested by Cui and Blockley (1990), who introduced interval probabilities based on probability axioms maintaining the additive condition of

Equation 4.104. For an event A that represents a proposition on a universal set X, the probability measure for A is given by

$$P(A) = [P_L(A), P_R(A)] \qquad (4.151)$$

where $P_L(A)$ and $P_R(A)$ are the upper (left) and lower (right) estimates of the probability of A, $P(A)$. According to Equation 4.151, the probability of A falls in this range as follows:

$$P_L(A) \leq P(A) \leq P_R(A) \qquad (4.152)$$

The probability of the complement of A can be computed as follows:

$$1 - P_R(A) \leq P(\overline{A}) \leq 1 - P_L(A) \qquad (4.153)$$

The interval probability can be interpreted as a measure of belief in having a true proposition A as follows:

$$P(A) = [0,0] \text{ represents a belief that } A \text{ is certainly false} \\ \text{or not dependable} \qquad (4.154)$$

$$P(A) = [1,1] \text{ represents a belief that } A \text{ is certainly true} \\ \text{or dependable} \qquad (4.155)$$

$$P(A) = [0,1] \text{ represents a belief that } A \text{ is unknown} \qquad (4.156)$$

The use of the term *belief* in Equations 4.154b to 4.156 should not be confused with the belief measure provided in the theory of evidence. Hall et al. (1998) provide an example application of interval probabilities.

4.3.4.11 Interval cumulative distribution functions

Probabilistic models are effective in expressing uncertainties in various variables that appear in engineering and scientific problems. Such models can be viewed as certain representations of uncertainty that demand knowledge of underlying distributions, parameters, and/or a lot of data. Systems that are represented by these models might not be known fully to the levels demanded by the models, hence the need of methods to deal with limited or incomplete information. Analysts commonly encounter situations where data are not available, limited, or available in intervals only. This section provides methods that were selected or developed to deal with such situations. The section covers three cases as follow: (1) uncertain parameters of a known probability distribution, (2) an uncertain probability distribution for known parameters, and (3) uncertain parameters and probability distribution due to limited data. These three cases are discussed with illustrative examples.

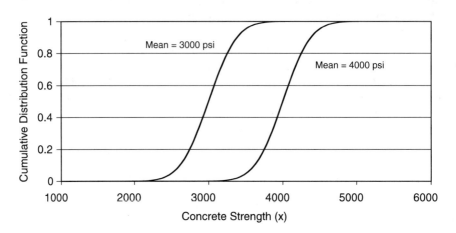

Figure 4.19 Normal cumulative distribution function using an interval mean.

For some random variables, the distribution type might be known from historical information; however the parameters relevant to a problem under consideration might not be known and can be only subjectively assessed using intervals or fuzzy numbers. The presentation herein is provided for interval parameters but can be easily extended to fuzzy parameters expressed as fuzzy numbers using the α-cut concept. If we consider a concrete structural member with an unknown strength, the following state of knowledge can be used to demonstrate the construction of an interval-based distribution:

Normal probability distribution
Mean value = [3000, 4000] psi
Standard deviation = 300 psi

The bounds of the cumulative distribution function ($F_X(x)$) are shown in Figure 4.19 based on evaluating the following integral:

$$F_X(x) = \int_{-\infty}^{x} \frac{1}{\sigma\sqrt{2}} \exp\left[-\frac{1}{2}\left(\frac{x-\mu}{\sigma}\right)^2\right] \sigma dx \qquad (4.157)$$

where μ = mean, and σ = standard deviation. Another case is shown in Figure 4.20 using the following assumptions:

Normal probability distribution
Mean value = [3000, 4000] psi
Standard deviation = [300, 400] psi, i.e., coefficient of variation = 0.10

Chapter four: Expressing and modeling expert opinions 191

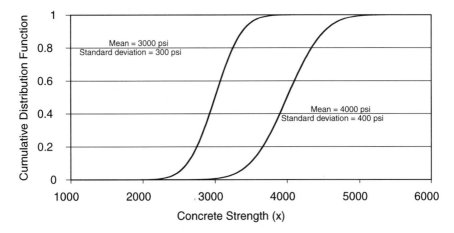

Figure 4.20 Normal cumulative distribution function using interval mean and standard deviation.

Some random variables might have known moments with unknown or uncertain distribution types. If data are available, one could use hypothesis testing to select a distribution that best fits the data (Ayyub and McCuen, 1997). However, data might not be available requiring the use of a bounding method. In this case a short list of distributions can be subjectively identified. The cumulative distribution functions based on the known parameters can be determined, and a range on possible values of the cumulative distribution function can be assessed.

Some random variables might have limited data that are not sufficient to construct a histogram and select a probability distribution. In this case, the Kolmogorov-Smirnov (KS) one-sample method can be used to construct a confidence interval on the cumulative distribution function. The KS method, as described by Ayyub and McCuen (1997), constructs a sample cumulative distribution function as follows:

$$F_S(x) = \begin{cases} 0 & \text{for } x < x_1 \\ \dfrac{i}{n} & \text{for } x_i \leq x < x_{i+1} \\ 1 & \text{for } x \geq x_n \end{cases} \quad (4.158)$$

where x_i = ith largest value based on rank-ordering the sample values from the smallest (x_1) to the largest (x_n) for a sample of size n, and $F_S(x)$ = the sample cumulative distribution function. The KS method provides tabulated limits on the maximum deviation between the sample cumulative distribution function and an acceptable model for the cumulative distribution function. These tabulated limits correspond to various sample sizes and significance levels, i.e., one minus the confidence level defined as the conditional probability of accepting a model given it is an incorrect model. Table 4.9 shows critical values for the KS method as a function of sample sizes and the significance levels. The following set of 5 measurements of a water quality parameter in ppm: {47, 53, 61, 57, 65} can be used to construct KS bounds on a cumulative distribution function. If a level of significance of 5 percent is used, a sample cumulative distribution function and bounds can be computed using Equation 4.158 and Table 4.9. Table 4.10 shows the calculations for sample cumulative distribution function and the two bounds. For a 5 percent level of significance, the critical value is 0.56. The sample, left and right cumulative distribution functions are shown in Figure 4.21.

Table 4.9 Critical Values for the Kolmogorov-Smirnov Test

Sample Size	Level of Significance			
n	0.20	0.10	0.05	0.01
5	0.45	0.51	0.56	0.67
10	0.32	0.37	0.41	0.49
15	0.27	0.30	0.34	0.40
20	0.23	0.26	0.29	0.36
30	0.19	0.22	0.24	0.29
40	0.17	0.19	0.21	0.25
50	0.15	0.17	0.19	0.23
60	0.138	0.296	0.176	0.210
70	0.125	0.146	0.163	0.195
80	0.120	0.136	0.152	0.182
90	0.113	0.129	0.143	0.172
100	0.107	0.122	0.136	0.163
>50	$1.07/\sqrt{n}$	$1.22/\sqrt{n}$	$1.36/\sqrt{n}$	$1.63/\sqrt{n}$

Table 4.10 Left and Right Bounds Using the Kolmogorov-Smirnov Limits

Sorted Data Point Rank i	Sample Value x	Sample CDF	Right CDF	Left CDF
0	20	0	0	0.56
1	47	0	0	0.56
1	47	0.2	0	0.76
2	53	0.2	0	0.76
2	53	0.4	0	0.96
3	57	0.4	0	0.96
3	57	0.6	0.04	1
4	61	0.6	0.04	1
4	61	0.8	0.24	1
5	65	0.8	0.24	1
5	65	1	0.44	1

Chapter four: Expressing and modeling expert opinions   193

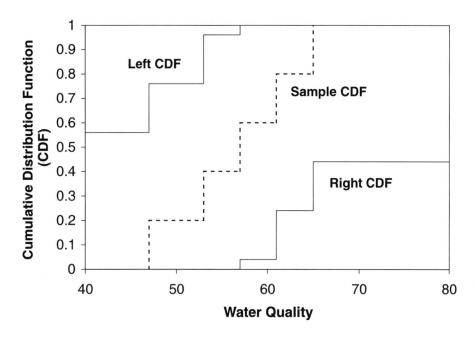

Figure 4.21 The Kolmogorov-Smirnov bounds on a cumulative distribution function.

4.3.4.12 *Probability bounds*

Probability bounds can be viewed as a mix of probability theory and interval analysis (Ferson et al., 1999). They have similar bases as interval probabilities and concepts covered in probabilistic analysis using limited or incomplete information. Probabilities in this case are uncertain and hence represented by probability bounds. Where random variables are used, cumulative distribution functions (CDF) offer a complete description of their probabilistic characteristics. Uncertainty in underlying parameters or limited knowledge about these variables result in the need to construct bounds on them.

For example, a random variable X might be known only to the extent of the minimum (e.g., $x = 50$) and maximum (e.g., $x = 70$) values that the variable could possibly take. The probability bounds for this random variable can be expressed in the form of CDF bounds as shown in Figure 4.22. The CDF bounds can be interpreted as the left and right limits on any possible CDF function that meets the constraint given by the minimum and maximum values of X. These CDF bounds can be denoted as $\underline{F}_X(x)$ and $\overline{F}_X(x)$ for left (or called lower on x) and right (or called upper on x) approximations of the CDF (i.e., F) of X. Increasing the level of information in this constraint results in reducing the gap between these bounds. For example, by adding a median value at $x = 60$ to the minimum-maximum constraint produces the CDF bounds of Figure 4.23. Figures 4.19, 4.20, and 4.21 offer additional examples of CDF bounds. Figures 4.19 and 4.20 can be approximated using interval

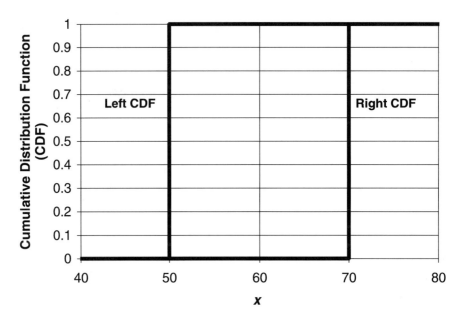

Figure 4.22 Bounds on a cumulative distribution function based on minimum and maximum values.

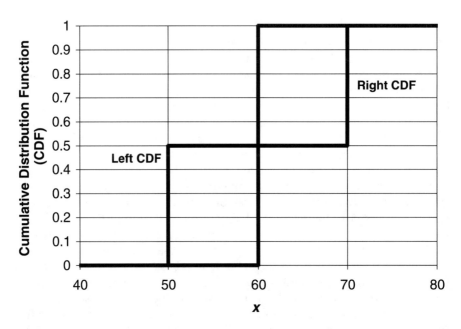

Figure 4.23 Bounds on a cumulative distribution function based on minimum, median, and maximum values.

Chapter four: Expressing and modeling expert opinions

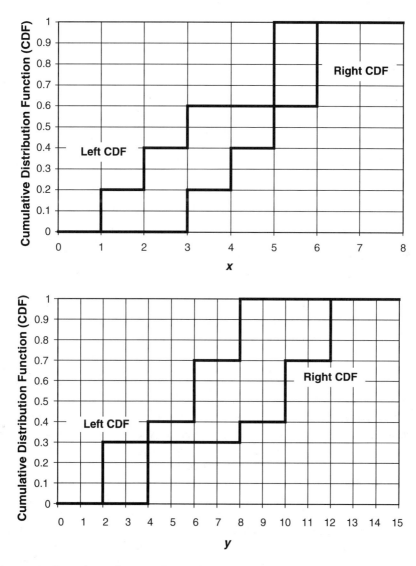

Figure 4.24 Bounds on the cumulative distribution functions of X and Y.

values on the underlying random variable X so that the resulting CDF bounds have the general step-function shapes provided in Figure 4.21.

To facilitate the probability calculus for these probability bounds, left and right approximations of the CDF of a random variable can be represented as step functions with an added restriction, for computational convenience, that the steps for both left and right functions occur at the same CDF values. Figure 4.24 provides examples of such CDF bounds for two random variables X and Y. These figures express uncertainty in the CDF. For example, Figure 4.24 provides the CDF bounds at $x = 3.5$ of [0.2, 0.6]. Also, the same figure expresses the uncertainty in the value of X at a given CDF value. For a CDF value (percentile value) of 0.90, the value of x belongs to the interval [5, 6].

Random variables defined by CDF bounds can be combined using arithmetic operations such as addition, subtraction, multiplication, and division; however information on underlying dependencies between the two random variables is needed in order to assess the combined result. Two cases are considered in this section as follows: (1) the case of an underlying independence between two random variables such as X and Y, and (2) the case of underlying, but unknown, dependencies between two random variables such as X and Y.

The underlying independence assumption for X and Y allows for computing, for example $X + Y$, by portioning the spaces of X and Y to convenient intervals, performing interval arithmetic on all the combinations of the Cartesian-space of X and Y, and computing the corresponding probabilities of the resulting intervals as the product of the respective pairs. The computational procedure is demonstrated using the random variable X and Y of Figure 4.24 to evaluate their addition, i.e., $X+Y$ as shown in Table 4.11. The left and right probability bounds of the CDF of the addition result Z can be evaluated as shown in Table 4.12. Table 4.12 was constructed by identifying the range of Z from Table 4.11, from 3 to 18 in general increments of one. Then for each Z value, such as z, the left bound was constructed as the cumulative sum of all interval probabilities for Z in Table 4.11 where the lower (left) limits of the intervals are less or equal to z. The right bound for Z can be constructed in a similar manner as the cumulative sum of interval probabilities from Table 4.11 where the upper (right) values of the intervals are less or equal to z. The resulting probability bounds of $Z = X + Y$ are shown in Figure 4.25. Other arithmetic operations such as subtraction, multiplication, and division can be performed in a similar manner to the above process for addition.

The case of an underlying, but unknown, dependency between two random variables such as X and Y requires arithmetic operations on X and Y to be conducted using a probability-bound convolution with the constraint that the sum of probabilities must be one. Frank et al. (1987), Nelson (1999), and Williamson and Downs (1990) provided the following probability bounds on $Z = X * Y$, where $* \in [+,-,\times,\div]$, for arithmetic operations with unknown dependencies between two random variables such as X and Y:

$$\underline{F}_{X+Y}(z) = \max_{\text{such that } z=u+v} \left\{ \max\left[\underline{F}_X(u) + \underline{F}_Y(v) - 1, 0\right] \right\} \qquad (4.159)$$

$$\overline{F}_{X+Y}(z) = \min_{\text{such that } z=u+v} \left\{ \min\left[\overline{F}_X(u) + \overline{F}_Y(v), 1\right] \right\} \qquad (4.160)$$

Chapter four: Expressing and modeling expert opinions

Table 4.11 Addition (Z = X+Y) Using CDF Bounds with Underlying Independence Expressed as Intervals with Probabilities

Intervals for Y and Their Probabilities	Intervals for X and Their Probabilities			
	P(1≤X<3) = 0.2	P(3≤X<4) = 0.2	P(4≤X<5) = 0.2	P(5≤X<6) = 0.4
P(2≤Y<4) = 0.3	P(3≤Z<7) = 0.06	P(5≤Z<8) = 0.06	P(6≤Z<9) = 0.06	P(7≤Z<10) = 0.12
P(4≤Y<8) = 0.1	P(5≤Z<11) = 0.02	P(7≤Z<12) = 0.02	P(8≤Z<13) = 0.02	P(9≤Z<14) = 0.04
P(8≤Y<10) = 0.3	P(9≤Z<13) = 0.06	P(11≤Z<14) = 0.06	P(12≤Z<15) = 0.06	P(13≤Z<16) = 0.12
P(10≤Y<12) = 0.3	P(11≤Z<15) = 0.06	P(13≤Z<16) = 0.06	P(14≤Z<17) = 0.06	P(15≤Z<18) = 0.12

Table 4.12 Probability Bounds for the Addition (Z=X+Y)

Addition Result of Z=X+Y	Left Bound	Right Bound
2	0	0
3	0.06	0
4	0.06	0
5	0.14	0
6	0.20	0
7	0.34	0.06
8	0.36	0.12
9	0.46	0.18
10	0.46	0.30
11	0.58	0.32
12	0.64	0.34
13	0.82	0.42
14	0.88	0.52
15	1.00	0.64
16	1.00	0.82
17	1.00	0.88
18	1.00	1.00
19	1.00	1.00
20	1.00	1.00

$$\underline{F}_{X-Y}(z) = \max_{\text{such that } z=u-v} \left\{ \max\left[\underline{F}_X(u) + \overline{F}_Y(-v), 0 \right] \right\} \quad (4.161)$$

$$\overline{F}_{X-Y}(z) = 1 + \min_{\text{such that } z=u-v} \left\{ \min\left[\overline{F}_X(u) - \underline{F}_Y(-v), 0 \right] \right\} \quad (4.162)$$

$$\underline{F}_{X\times Y}(z) = \max_{\text{such that } z=u\times v} \left\{ \max\left[\underline{F}_X(u) + \underline{F}_Y(v) - 1, 0 \right] \right\} \quad (4.163)$$

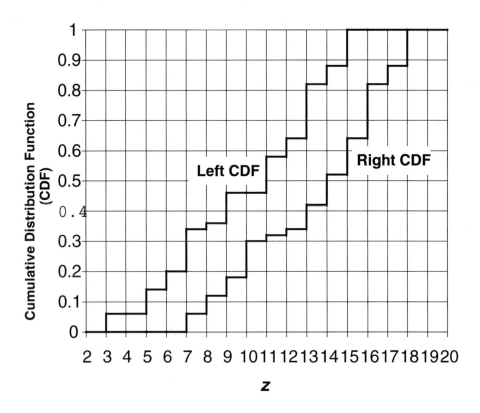

Figure 4.25 Bounds on the cumulative distribution functions of Z = X + Y.

$$\overline{F}_{X \times Y}(z) = \min_{\text{such that } z = u \times v} \left\{ \min\left[\overline{F}_X(u) + \overline{F}_Y(v), 1 \right] \right\} \quad (4.164)$$

$$\underline{F}_{X+Y}(z) = \max_{\text{such that } z = u+v} \left\{ \max\left[\underline{F}_X(u) - \overline{F}_Y(1/v), 0 \right] \right\} \quad (4.165)$$

$$\overline{F}_{X+Y}(z) = 1 + \min_{\text{such that } z = u+v} \left\{ \min\left[\overline{F}_X(u) - \underline{F}_Y(1/v), 0 \right] \right\} \quad (4.166)$$

Williamson and Downs (1990) showed that the above bounds hold for the arithmetic operation of addition and multiplication for both positive and negative X and Y, and for the arithmetic operation of subtraction and division for only positive X and Y. Regan et al. (2000) generalized Equations 4.159 to 4.166 to any arithmetic operation regardless of the sign of X and Y by using the interval mathematics of Equations 4.37 to 4.40 that, in the case of subtraction and division, combines the lower bound of one variable with the upper bound of another and vise versa as required by these equations. For two events A and B with given probabilities $P(A)$ and $P(B)$, the limits provided by Equations 4.159 to 4.166 are partially based on the conjunction and disjunction, called Frechet, inequalities as follow, respectively:

$$\text{Conjunction: } \max(0, P(A) + P(B) - 1) \le P(A \cap B) \le \min(P(A), P(B)) \tag{4.167}$$

$$\text{Disjunction: } \max(P(A), P(B)) \le P(A \cup B) \le \min(1, P(A) + P(B)) \tag{4.168}$$

Equations 4.167 and 4.168 usually result in wide limits, and their use for CDF functions can violate the constraint that the sum of probabilities must be one, whereas Equations 4.159 to 4.166 do not violate this constraint.

Regan et al. (2000) showed the equivalency of Equations 4.159 to 4.166 in propagating uncertainty to methods offered by Walley (1991) for imprecise probabilities, and Dempster-Shafer belief functions as provided by Yager (1986).

4.3.5 Possibility theory

Possibility theory and its monotone measures of necessity and possibility are based on crisp sets and the nonadditive properties of Equations 4.103 and 4.105 as described by Klir and Folger (1988), Klir and Wierman (1999), Dubois and Prade (1988), and De Cooman (1997). The possibility theory is a special case of the Dempster-Shafer theory of evidence and its monotone measures of belief and plausibility by requiring the underlying subsets of a universe X to be nested, i.e., $A_1 \subset A_2 \subset \ldots \subset X$. Nested subsets on X are called *chains*. Nested subsets for an evidence body result in minimal conflicts with each other; therefore, their belief and plausibility measures, called necessity and possibility measures respectively, in this case, are described to be *consonant*. An example of five nested sets (A_i) with 10 discrete elements (x_j) is shown in Figure 4.26. For nested subsets, the associated belief and plausibility measures, i.e., necessity and possibility measures, respectively, satisfy the following conditions:

$$Bel(A_1 \cap A_2) = \min[Bel(A_1), Bel(A_2)] \text{ for any } A_1 \text{ and } A_2 \in P_X \tag{4.169}$$

and

$$P(A_1 \cup A_2) = \max[Pl(A_1), Pl(A_2)] \text{ for any } A_1 \text{ and } A_2 \in P_X \tag{4.170}$$

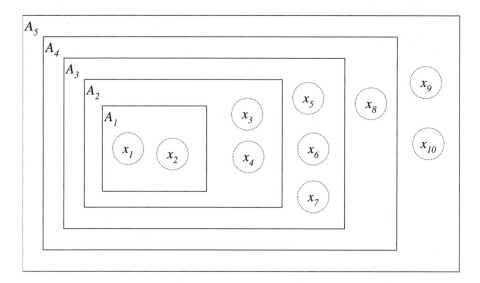

Figure 4.26 Nested sets and singletons for a possibility distribution.

A possibility distribution function $r(x)$ is defined as a mapping from the universal set X to the range $[0,1]$ according to the following equation:

$$r : X \to [0,1] \qquad (4.171)$$

where $\max(r(x)) = 1$. The possibility measure (Pos) for a subset $A_i \subseteq X$ can be uniquely determined based on $r(x)$ as follows:

$$Pos(A_i) = \max_{a\mu\, x \in A_i}(r(x)) \qquad (4.172)$$

The possibility measure is, therefore, a mapping from the power set of X to the interval $[0,1]$. The corresponding necessity (Nec) measure for a subset $A_i \subseteq X$ can be defined as

$$Nes(A_i) = 1 - Pos(\overline{A}_i) \qquad (4.173)$$

where $\overline{A}_i$ is the complement of A. Therefore, the necessity measure is also a mapping from the power set of X to the interval $[0,1]$.

The following properties of possibility and necessity measures are provided for any pairs of subsets, $A_i \subseteq X$ and $A_j \subseteq X$:

$$Pos(A_i \cup A_j) = \max[Pos(A_i), Pos(A_j)] \qquad (4.174)$$

$$Nes(A_i \cap A_j) = \min[Nec(A_i), Nec(A_j)] \qquad (4.175)$$

Chapter four: Expressing and modeling expert opinions

$$Pos(A_i \cap A_j) \leq \min[Pos(A_i), Pos(A_j)] \quad (4.176)$$

$$Nes(A_i \cup A_j) \geq \max[Nec(A_i), Nec(A_j)] \quad (4.177)$$

$$Nec(A_i) \leq Pos(A_i) \text{ for all } A_i \quad (4.178)$$

$$Pos(A_i) + Pos(\overline{A_i}) \geq 1 \text{ for all } A_i \quad (4.179)$$

$$Nec(A_i) + Nec(\overline{A_i}) \leq 1 \text{ for all } A_i \quad (4.180)$$

$$\max[Pos(A_i), Pos(\overline{A_i})] = 1 \text{ for all } A_i \quad (4.181)$$

$$\min[Nec(A_i), Nec(\overline{A_i})] = 0 \text{ for all } A_i \quad (4.182)$$

$$Pos(A_i) < 1 \Rightarrow Nec(A_i) = 0 \text{ for all } A_i \quad (4.183)$$

$$Nec(A_i) > 0 \Rightarrow Pos(A_i) = 1 \text{ for all } A_i \quad (4.184)$$

The nested structure of a family of sets, i.e., $A_1 \subset A_2 \subset \ldots \subset X$, is compatible with the α-cuts of convex fuzzy sets, making fuzzy set interpretation of possibility theory logical. Klir and Folger (1988), Klir and Wierman (1999), Dubois and Prade (1988), and De Cooman (1997) provide additional details on possibility theory and its applications.

4.4   Exercise problems

Problem 4.1 A construction manager needs to procure building materials for the construction of an office building. The following sources are identified:

Material Type	Sources
concrete	Sources A and B
reinforcing steel	Sources C and D
timber	Sources D and E
structural steel	Sources C and F
hardware	Sources F, G, and H

Define the sample space of all possible combinations of sources supplying the construction project; assume that each material type can be procured from one source only, but a source may supply more than one material type at the same time.

Problem 4.2 A construction tower crane can operate up to a height H of 300 ft, a range (radius) R of 50 ft, and an angle ϕ of $\pm 90°$ in a horizontal plane. Sketch the sample space of operation of the crane. Sketch the following events:

Event	Definition
A	$30<H<80$ and $R<30$
B	$H>50$ and $0°<\phi<50°$
C	$H<40$ and $R>60$
D	$H>80$ and $-30°<\phi<50°$

Problem 4.3 Construct Venn diagrams for each of the following:

1. A deck of playing cards.
2. The roll of a die.
3. Letter grades on a test assuming equal probabilities for each grade.
4. Letter grades on a test assuming the following probabilities:

    A: 15%; B: 25%; C: 30%; D: 20%; and F: 10%.

5. Options at an intersection with the following probabilities:

    Left turn: 20%; Straight ahead 40%; Right turn: 25%; U-turn: 10%; and Remain stopped: 5%.

Problem 4.4 For the data and events of Problem 4.2, sketch the following events:

$A \cup B, A \cap B, C \cup D, C \cap D, A \cup C, A \cup (B \cap C), \overline{A},$ and $\overline{A} \cap B$

Problem 4.5 The traffic that makes a left turn at an intersection consists of two types of vehicles, types A and B. A type A vehicle is twice the length of type B. The left-turn lane can accommodate 8 vehicles of type B, 4 of type A, or combinations of A and B. Define the sample space of all possible combinations of vehicles waiting for a left turn at the intersection left turn lane. Also define the following events: (1) at least one vehicle of type A waiting for left turn, (2) two vehicles of type B waiting for left turn, and (3) exactly one of type A and one of type B waiting for a left turn.

Problem 4.6 Construct a Venn diagram for a deck of playing cards (4 suits, 13 cards per suit). Show the following events:

1. A = all diamonds and all aces;
2. B = all face cards;
3. C = the intersection of red cards and face cards;
4. D = the union of black cards and cards with values of 4 or smaller.

Problem 4.7 Using the α-cut concept, compute the intersection, union, and complements of the following fuzzy sets:

1. triangular sets defined as A = [10, 15, 20] and B = [15, 18, 22]; and
2. trapezoidal sets defined as C = [10, 12, 18, 20] and D = [15, 18, 20, 22].

Plot your results.

Chapter four: Expressing and modeling expert opinions   203

Problem 4.8 Using the alpha-cut concept, compute the intersection, union and complements of the triangular set defined as $A = [10, 15, 20]$ and trapezoidal set defined as $B = [10, 12, 18, 20]$. Plot your results.

Problem 4.9 Using the alpha-cut concept, evaluate the following operations on the triangular set defined as $A = [10, 15, 20]$ and trapezoidal set defined as $B = [10, 12, 18, 20]$, and plot your results:

1. $A + B$;
2. $A - B$;
3. $A \times B$;
4. A / B;
5. $A + B$, with the constraint that $a = b$ where $a \in A$ and $b \in B$;
6. $A - B$, with the constraint that $a = b$ where $a \in A$ and $b \in B$;
7. $A \times B$, with the constraint that $a = b$ where $a \in A$ and $b \in B$; and
8. A / B, with the constraint that $a = b$ where $a \in A$ and $b \in B$.

Problem 4.10 Using the α-cut concept, evaluate the following operations on the triangular set defined as $A = [-10, 15, 20]$ and trapezoidal set defined as $B = [10, 12, 18, 20]$, and plot your results:

1. $A + B$;
2. $A - B$;
3. $A \times B$;
4. A / B;
5. $A + B$, with the constraint that $a = b$ where $a \in A$ and $b \in B$;
6. $A - B$, with the constraint that $a = b$ where $a \in A$ and $b \in B$;
7. $A \times B$, with the constraint that $a = b$ where $a \in A$ and $b \in B$; and
8. A / B, with the constraint that $a = b$ where $a \in A$ and $b \in B$.

Problem 4.11 Construct a fuzzy relation $A \times B$ based on the triangular set defined as $A = [10, 15, 20]$ and the trapezoidal set defined as $B = [10, 12, 18, 20]$, and plot your results.

Problem 4.12 Develop an equation to compute an average of n fuzzy data points that are of

1. triangular fuzzy membership functions;
2. trapezoidal fuzzy membership functions; and
3. mixed triangular and trapezoidal fuzzy membership functions.

Problem 4.13 Building on Problem 4.12, develop equations needed to compute variance of n fuzzy data points that are of

1. triangular fuzzy membership functions;
2. trapezoidal fuzzy membership functions; and
3. mixed triangular and trapezoidal fuzzy membership functions.

What are the limitations of your equations?

Problem 4.14 Provide an example unknown set, and function that can be represented by a rough set and a rough function, respectively.

Problem 4.15 Assume that the x's represent users or intruders of an information network with specific nested characteristics such as x_1 = a legitimate user from an authorized IP address, x_2 = an intruder from an authorized IP address or an intruder from an IP address that is masked by an authorized IP address, x_3 = a legitimate user with an authorized access protocol (username and password), and x_4 = an intruder from an authorized IP access protocol. The nested structure of Figure 4.26 can be used to construct evidence gathering methods and their probabilities of affirmative detection, false detection, affirmative nondetection, and false nondetection. These probabilities can be constructed as basic assignments to meet the requirements of the theory of evidence. They are then used to compute belief and plausibility measures for having any of the events $A_1 \ldots A_5$ of Figure 4.26. Develop the mathematical formulation needed for this application with an illustrative example.

Problem 4.16 Two judges classified a case to three possible motivations as provided in the following table in the form of basic assignments m_1 and m_2:

Subset (i.e., Motivation)	Judge 1 m_1	Judge 2 m_2
Greed (G)	0.05	0.15
Love (L)	0.10	0.05
Self defense (S)	0.15	0.05
$G \cup L$	0.25	0.15
$G \cup S$	0.15	0.20
$L \cup S$	0.05	0.30
$G \cup L \cup S$	Not provided	Not provided

Compute the belief measures for judges 1 and 2. Compute the basic assignment for the combined judgment and the corresponding belief measure.

Problem 4.17 Two experts classified three possible causes for a bird species population decline, as provided in the following table, in the form of basic assignments m_1 and m_2:

Subset (i.e., Cause)	Expert 1 m_1	Expert 2 m_2
Changes in land use (C)	0.10	0.15
Hunting (H)	0.15	0.15
Disease (D)	0.15	0.15
$C \cup H$	0.25	0.15
$C \cup D$	0.15	0.10
$H \cup D$	0.15	0.20
$C \cup H \cup D$	Not provided	Not provided

Chapter four: Expressing and modeling expert opinions

Compute the belief measures for experts 1 and 2. Compute the basic assignment for the combined judgment and the corresponding belief measure.

Problem 4.18 Using Table 4.6, compute the joint of probability of the two events A and B that have the following probabilities:

Probability (A) = likely
Probability (B) = Seldom

Treat the above probabilities as fuzzy sets and use the α-cut method to compute the probability of A and B, assuming that they are independent events. Express your result using a fuzzy set and linguistically basing it on Table 4.6. What are the limitations of such a hybrid use of linguistic probabilities and fuzzy sets?

Problem 4.19 The accident probability at a new intersection is of interest to a traffic engineer. The engineer subjectively estimated the weekly accident probability as follows:

Weekly Accident Probability	Subjective Probability of Accident Probability
0.1	0.3
0.2	0.4
0.4	0.2
0.6	0.05
0.8	0.04
0.9	0.01

Solve the following:

1. What is the average accident probability based on the prior information?
2. Given an accident in the first week of traffic, update the distribution of the accident probability.
3. What is the new average accident probability based on the posterior information?
4. Given accidents in the first and second weeks and no accidents in the third week of traffic, update the distribution of the accident probability.
5. What is the average accident probability after the second week?
6. Given no additional accidents for the weeks 4, 5, 6, 7, 8, 9, and 10, update the distribution and the average accident probability. Plot your results.

Problem 4.20 Plot an interval distribution for the following random variable:

Logormal probability distribution
Mean value = [3000, 4000] psi
Standard deviation = [300, 300]

Problem 4.21 Plot an interval distribution for the following random variable:

Logormal probability distribution
Mean value = [3000, 4000] psi
Standard deviation = [300, 400] psi, i.e., coefficient of variation = 0.10

Problem 4.22 Plot an interval distribution for the following random variable:

Exponential probability distribution
Mean value = [3000, 4000]

Problem 4.23 Redo the example shown in Tables 4.11 and 4.12, and Figure 4.25 for the multiplication operation instead of the addition of X and Y, i.e., $Z = X \times Y$.

Problem 4.24 Redo the example shown in Tables 4.11 and 4.12, and Figure 4.25 for the subtraction operation instead of the addition of X and Y, i.e., $Z = X - Y$.

Problem 4.25 Redo the example shown in Tables 4.11 and 4.12, and Figure 4.25 for the division operation instead of the addition of X and Y, i.e., $Z = X \div Y$.

chapter five

Consensus and aggregating expert opinions

Contents

- 5.1. Introduction 208
- 5.2. Methods of scoring of expert opinions 208
  - 5.2.1. Self scoring 208
  - 5.2.2. Collective scoring 209
- 5.3. Uncertainty measures 209
  - 5.3.1. Types of uncertainty measures 209
  - 5.3.2. Nonspecificity measures 209
  - 5.3.3. Entropy-like measures 211
    - 5.3.3.1. Shannon entropy for probability theory 212
    - 5.3.3.2. Discrepancy measure 212
    - 5.3.3.3. Entropy measures for evidence theory 213
  - 5.3.4. Fuzziness measure 214
  - 5.3.5. Other measures 214
- 5.4. Combining expert opinions 214
  - 5.4.1. Consensus combination of opinions 215
  - 5.4.2. Percentiles for combining opinions 215
  - 5.4.3. Weighted combinations of opinions 215
  - 5.4.4. Uncertainty-based criteria for combining expert opinions 218
    - 5.4.4.1. Minimum uncertainty criterion 219
    - 5.4.4.2. Maximum uncertainty criterion 219
    - 5.4.4.3. Uncertainty invariance criterion 221
  - 5.4.5. Opinion aggregation using interval analysis and fuzzy arithmetic 221
  - 5.4.6. Opinion aggregation using Dempster's rule of combination 222
  - 5.4.7. Demonstrative examples of aggregating expert opinions 222
    - 5.4.7.1. Aggregation of expert opinions 222
    - 5.4.7.2. Failure classification 227
- 5.5. Exercise problems 230

5.1 Introduction

Chapter 1 provides background materials on the definitions of knowledge, ignorance, information, uncertainty, and expert opinions. Experts render subjective opinions based on existing knowledge and information available to them. Consequently, expert opinions can be viewed as preliminary propositions with claims that are not fully justified or are justified with adequate reliability but are not infallible. In other words, expert opinions are seeds of propositional knowledge that do not meet one or more of the conditions required for the justified true belief (JTB) within the framework of the reliability theory of knowledge. Despite the uncertainties associated with expert opinions, we consider them valuable since they might lead to knowledge expansion or evolution. In engineering and science, decisions commonly need to be made based on these opinions. These decisions can be considered risky, especially in cases of expert opinions with a lot of uncertainty, including conflicting or confusing opinions. Such decision situations structured using preliminary propositions can lead to adverse outcomes as some of the propositions might be proved false. The relationships among knowledge, information, opinions, and evolutionary epistemology are schematically shown in Figure 1.6. The dialectic processes include communication methods such as languages, visual and audio formats, and other forms. It is common in situations of high uncertainty to utilize multiple experts and assess the reliability of the experts (i.e., the sources of the opinions) using scoring methods. Also, once the expert opinions are gathered, the uncertainty contents in these opinions individually and collectively need to be measured, and a final opinion needs to be produced through consensus or aggregation.

The objective of this chapter is to present methods for assessing or scoring expert opinions, measuring the uncertainty contents in individual opinions, aggregated or combined opinions, and selecting an optimal opinion. The methods presented are based on developments in expert opinion elicitation and uncertainty-based information of information science.

5.2 Methods of scoring of expert opinions

Scoring methods can be used to assess the information reliability (or quality) provided by experts through an expert-opinion elicitation process. The resulting scores can then be used for various applications, for example, to determine weight factors for combining expert opinions if needed. Scoring methods depend on the experts to assess either the reliability of their opinions or those of other experts; however, such approaches are like having a fox guarding a henhouse.

5.2.1 Self scoring

According to this method, each expert provides a self-assessment in the form of a confidence level for each probability or answer provided for an issue.

The primary disadvantages of this method are bias and overconfidence that can result in inaccurate self assessments. The method has the advantage of simplicity that can be viewed as commensurate to the nature of use of expert-opinion elicitation.

5.2.2 Collective scoring

According to this method, each expert provides assessments of other experts' opinions in the form of confidence levels in their provided probabilities or answers related to an issue. The primary disadvantages of this method are bias and nonreproducibility. The bias in this case can be in an opposite direction to the bias that might result from the self-scoring method — for example, an underestimation versus an overestimation.

5.3 Uncertainty measures

5.3.1 Types of uncertainty measures

We have presented several uncertainty types in Chapter 4 that were modeled using various theories, such as probability, possibility, and evidence. Engineers and scientists would find great benefit in having measures to quantify these uncertainties. These measures would be similar to measuring physical quantities, such as temperature, pressure, or dimensions; however, they are unique in that they measure conceived or abstract notions rather than a physical quantity. These uncertainty measures can be defined to be nonnegative real numbers and should be inversely proportional to the strength and consistency in evidence as expressed in the theory employed, i.e., the stronger and more consistent the evidence, the smaller the amount of uncertainty (Klir and Wierman, 1999). Such uncertainty measures should be suitable to assess opinions rendered by one expert on some issue of interest, or opinions rendered by several experts on the same issue.

The area of uncertainty measures is not fully developed, and so is an active research area (Klir and Wierman, 1999). Uncertainty measures were suggested for some uncertainty types and theories covered in this chapter. This section limits itself to three relatively mature uncertainty types: (1) nonspecificity that results from imprecision connected with set sizes — cardinalities — and can be represented by the Hartley-like measure; (2) likelihood that results from various basic assignments represented by Entropy-like uncertainty measures; and (3) fuzziness as a result of vagueness. Engineering and science problems and expert opinions can simultaneously contain these uncertainty types and others.

5.3.2 Nonspecificity measures

A fundamental uncertainty type stems from lack of specificity as a result of providing several alternatives, with one alternative being the true one. This uncertainty type vanishes, and complete certainty is achieved, when one alternative is presented. Therefore, the nonspecificity uncertainty type results from having imprecision due to alternative sets that have cardinalities greater than one. This fundamental measure can therefore be defined for a finite set of all possible alternatives, i.e., universal space X of alternatives under consideration, with only one of the alternatives being correct, although the correct alternative is unknown to us. However, we know based on all available evidence that the true alternative is in a subset A of X. In this case, only the alternatives that belong to A are considered as possible candidates for this true alternative. A measure of the amount of uncertainty associated with any finite set A of possible alternatives can be defined using the Hartley measure (H) as follows (Hartley, 1928):

$$H(A) = \log_2(|A|) \tag{5.1}$$

where $|A|$ is the cardinality of A, and $\log_2$ is the logarithm to the base 2 resulting in a measurement unit in bits. A bit is a single digit in a binary number system and can be viewed as a unit of information equal to the amount of information obtained by learning or by resolving which of two equally likely events has occurred. In computer language, *bits* form the basis of a *byte*, a string of binary digits (bits), usually eight, operated on as a basic unit by a digital computer. The logarithm ($\log_2$) of i as given by

$$\log_2(i) = x \tag{5.2a}$$

is the power to which the base, in this case 2, must be raised to obtain i as provided by

$$2^x = i \tag{5.2b}$$

Chapter 4 provided cases that involve classifying an element to a family of subsets using the theory of evidence. The nonspecificity in evidence can be constructed be extending the Hartley measure to each subset, and computing a weighted sum of all the resulting measures of the subsets using the basic assignment as weight factors. The nonspecificity measure can therefore be defined for a basic assignment m for a family of subsets, $A_1, A_2, \ldots, A_n \in P_X$, according to the theory of evidence (H_e) as follows:

$$H_e(m) = \sum_{i=1}^{n} m(A_i) \log_2(|A_i|) \tag{5.3}$$

Equation 5.3 provides an assessment of the nonspecificity in evidence. The nonspecificity in an evidence results from associating the basic assignment values to subsets that each can contain more than one element. This uncertainty can be eliminated by making the assignments m singletons, i.e., individual elements of X. Equation 5.3 becomes zero by having a body of evidence of singletons, i.e., $|A_i| = 1$ for all i. The uncertainty herein is due to nonspecificity within each subset A_i, and is not due to having more than one subset in the family of sets. A nonspecificity uncertainty that is associated with more than one subset in the family of sets can be defined within the framework of possibility theory of Section 4.3.5. The nonspecificity measure H_p for a basic assignment m for a family of nested subsets, A_1, A_2, ..., $A_n \in P_X$, that are singletons based on a possibility distribution $r(x)$ can be defined as follows:

$$H_p(r) = \sum_{i=2}^{n}\left[(r(x_i) - r(x_{i+1}))\log_2(i)\right] \tag{5.4}$$

where $r(x_{n+1}) = 0$ by convention, $H_p(r)$ is commonly referred to as the *U-uncertainty*. The possibility distribution is defined for a finite discrete universal set $X = \{x_1, x_2, ..., x_n\}$ as $r = \{r(x_1), r(x_2), ..., r(x_n)\}$ with $r(x_i) \geq r(x_{i+1})$.

Higashi and Klir (1983) provided the following nonspecificity measure, the U-uncertainty, for a normal fuzzy set (A):

$$U(A) = \int_0^1 \log_2(|{}^\alpha A|) d\alpha \tag{5.5}$$

where $|{}^\alpha A|$ = the cardinality of the α-cut of A.

The Hartley measure can be extended to deal with fuzzy sets and infinite sets. Klir and Folger (1988) and Klir and Wierman (1999) provide additional information on other Hartley measures and Hartley-like measures.

5.3.3 Entropy-like measures

Experts can be asked to provide a probability mass function that is associated with all possible values for an issue of interest such as occurrence probability of events. Assuming that there are n possible values, the probability assignment (P) by an expert can be expressed as p_i, $i = 1, 2, ..., n$. The uncertainty in this case has two aspects: (1) nonspecificity due to the existence of more than one possible outcome, and (2) conflict as described by the likelihood distribution provided by the probability mass function. The Hartley measure as provided in the previous section is well suited for the former aspect, but it does not cover the latter. The Shannon entropy was developed to measure the conflict uncertainty associated with a likelihood assignment for finite sets (Shannon, 1948). It was extended to measure uncertainty based on basic assignments in evidence theory. The basic Shannon entropy and its extensions to evidence theory are provided in this section. Many attempts were made to generalize the Shannon entropy in evidence theory and other theories, but they have had limited success. Suggested models in this area are called entropy-like measures and are not fully mature for the purposes of this book.

5.3.3.1 Shannon entropy for probability theory

Shannon (1948) provided an uncertainty measure for conflict that arises from a probability mass function that is commonly known as the *entropy measure* or the *Shannon entropy measure*. The entropy measure $S(P)$ measure is given by

$$S(P) = -\sum_{i=1}^{n} p_i \log_2(p_i) \quad (5.6)$$

This entropy measure takes on values from 0 to 1. Its value is zero if $p_i = 1$ and $n = 1$, and it is one for equally likely outcomes of $p_i = 1/n$ for all i. Klir and Folger (1988), and Klir and Wierman (1999) provide more details on this measure. For a continuous random variable, the entropy is called the Boltzman (1894) entropy as given by (Harr, 1987):

$$B(f) = -\int_{a}^{b} f_X(x) \log_2(f_X(x)) dx \quad (5.7)$$

where a = lower limit, b = upper limit, and f = probability density function.

5.3.3.2 Discrepancy measure

An expert can be used to estimate a probability mass function (P) expressed as p_i, $i = 1, 2, \ldots, n$. This function is an estimate of a true, yet unknown, probability mass function (S) expressed as s_i, $i = 1, 2, \ldots, n$. The discrepancy measure is between the true and provided probability values as given by

$$S_D(S,P) = -\sum_{i=2}^{n} s_i \log_2\left(\frac{s_i}{p_i}\right) \quad (5.8)$$

This discrepancy measure (S_D) is based on the Shannon entropy measure. The discrepancy measure can be used to obtain assessments of opinions obtained from a set of experts with equal circumstances and conditions, although equal circumstances and conditions might not be attainable. The discrepancy measure provides an assessment of the degree of surprise that someone would experience if an estimate p_i, $i = 1, 2, \ldots, n$ is obtained whereas the real values are s_i, $i = 1, 2, \ldots, n$ (Cooke, 1991).

5.3.3.3 Entropy measures for evidence theory

5.3.3.3.1 Measure of dissonance. Dissonance is a state of contradiction between claims, beliefs, or interests (Yager, 1983). The measure of dissonance, D, can be defined based on evidence theory as follows:

$$D(m) = -\sum_{i=1}^{n} m(A_i)\log_2(Pl(A_i)) \quad (5.9)$$

where

$m(A_i) > 0$;
$\{A_1, A_2, \ldots, A_n\}$ = a family set of subsets — focal points — that contains some or all elements of the universal set X;
$m(A_i)$ = a basic assignment which is interpreted either as the degree of evidence supporting the claim that a specific element belongs to the subset A_i but not to any special subset of A_i, or as the degree of belief that such a claim is warranted;

$\sum_{i=1}^{n} m(A_i) = 1$; and

$Pl(A_i)$ = plausibility measure which represents the total evidence or belief that the element of concern belongs to the set A_i or to any other sets that intersect with A_i as provided by Equation 4.119.

Klir and Folger (1988) extended to the dissonance measure to fuzzy sets A_i using possibility theory.

5.3.3.3.2 Measure of confusion. The measure of confusion characterizes the multitude of subsets supported by evidence as well as the uniformity of the distribution of strength of evidence among the subsets. The greater the number of subsets involved and the more uniform the distribution, the more confusing the presentation of evidence (Klir and Folger, 1988). The measure of confusion, C, is defined as

$$C(m) = -\sum_{i=1}^{n} m(A_i)\log_2(Bel(A_i)) \quad (5.10)$$

where $Bel(A_i)$ = belief measure that represents the total evidence or belief that the element of concern belongs to the subset A_i as well as to the various special subsets of A_i as provided by Equation 4.118. Klir and Folger (1988) extend to the confusion measure to fuzzy sets A_i using possibility theory.

5.3.4 Fuzziness measure

Fuzziness as represented by fuzzy set theory results from uncertainty in belonging to a set or notion. Fuzzy sets are sets that have fuzzy boundaries. For a given fuzzy set A, each element x of the universal set X has a membership value $\mu_A(x)$ to A that represents the degree of belief that x belongs to A or can be viewed as a measure of compatibility between x and the notion of A. The membership value is in the range [0,1] as described in Section 4.2.3. The vagueness uncertainty type can be assessed based on the fuzzy boundaries of the set A. A set A is considered fuzzier than set B if the boundary of A becomes wider than the boundaries of B in defining the transition from belonging to nonbelonging to the set. In addition, a fuzzy set has a unique property of having a nonempty intersection with its complement. Yager (1979 and 1980b) use these interpretations of fuzziness in suggesting a measure for it as follows:

$$f(A) = |X| - \sum_{x \in X} |\mu_A(x) - \mu_{\overline{A}}(x)| \tag{5.11}$$

f = fuzziness measure of a fuzzy set A; X = universal set; $\overline{A}$ = complement of A; $\mu_A(x)$ is membership value of x to A. Using the definition of the complement provided by Equation 4.52, Equation 5.11 can be written as

$$f(A) = |X| - \sum_{x \in X} |2\mu_A(x) - 1| \tag{5.12}$$

The fuzziness measure becomes zero as the set becomes crisp with $\mu_A(x)$ taking only values of zeros and ones and reaches maximum at $\mu_A(x) = 0.5$.

5.3.5 Other measures

Klir and Wierman (1999) describe other uncertainty measures, such as the discord measure and strife measure based on evidence theory and possibility theory. They also provide methods to aggregate uncertainty based on evidence theory, fuzzy entropy based on fuzzy set theory, and generalized measure based on fuzzified evidence theory.

5.4 Combining expert opinions

In some applications, expert opinions in the form of subjective probabilities of an event need to be combined into a single value and perhaps intervals for their use in probabilistic and risk analyses. Cooke (1991) and Rowe (1992) provide a summary of methods for combining expert opinions. The methods can be classified into consensus methods and mathematical methods (Clemen, 1989; Ferrell, 1985). The mathematical methods can be based on assigning equal weights or different weights to the experts. This section provides a summary of methods for combining expert opinions.

5.4.1 Consensus combination of opinions

A consensus combination of opinion is obtained through a facilitated discussion among the experts to some agreeable common values with perhaps a confidence interval or outer quartile values. The primary shortcomings of this method are (1) socially reinforced irrelevance or conformity within a group, (2) dominance of strong-minded or strident individuals, (3) group motive of quickly reaching an agreement, and (4) group-reinforced bias due to common background of group members. The facilitator of an expert-opinion elicitation session should play a major role in reducing group pressure, individual dominance, and biases.

5.4.2 Percentiles for combining opinions

A p-percentile value (x_p) for a random variable based on a sample can be defined as the value of the parameter such that $p\%$ of the data is less than or equal to x_p. On the basis of this definition, the median value is considered to be the 50-percentile value. Aggregating the opinions of experts can be based on computing the 25, 50, and 75 percentile values of the gathered opinions. The computation of these values depends on the number of experts providing opinions. Table 4.7 provides a summary of the needed equations for 4 to 20 experts. For example, 7 experts provided the following subjective probability of an event sorted in decreasing order:

$$\text{Probabilities} = \{1.0\text{E-}02, 5.0\text{E-}03, 5.0\text{E-}03, 1.0\text{E-}03, 1.0\text{E-}03, 5.0\text{E-}04, 1.0\text{E-}04\} \quad (5.13)$$

The median and arithmetic quartile points according to Table 4.7 are respectively given by

$$25 \text{ percentile} = 5.0\text{E-}03 \quad (5.14a)$$

$$50 \text{ percentile (median)} = 1.0\text{E-}03 \quad (5.14b)$$

$$75 \text{ percentile} = 7.5\text{E-}04 \quad (5.14c)$$

5.4.3 Weighted combinations of opinions

French (1985) and Genest and Zidek (1986) provide summaries of various methods for combining probabilities and example uses. For E experts with the ith expert providing a vector of n probability values, $p_{1i}, p_{2i}, \ldots, p_{ni}$, for sample space outcomes $A_1, A_2, \ldots, A_n$, the E expert opinions can be combined using weight factors $w_1, w_2, \ldots, w_E$, that sum up to one, using one of the following selected methods:

1. **Weighted arithmetic average:**
The weighted arithmetic mean for outcome j can be computed as follows:

$$\text{Weighted arithmetic mean for outcome } j = M_1(j) = \sum_{i=1}^{E} w_i p_{ji} \quad (5.15)$$

The weighted arithmetic means are then normalized using their total to obtain the 1-norm probability for an outcome for each outcome as follows:

$$\text{1-norm probability for outcome } j = P_1(j) = \frac{M_1(j)}{\sum_{k=1}^{n} M_1(k)} \quad (5.16)$$

2. **Weighted geometric average:**
The weighted geometric mean for outcome j can be computed as follows:

$$\text{Weighted geometric mean for outcome } j = M_0(j) = \prod_{i=1}^{E} (p_{ji})^{w_i} \quad (5.17)$$

The weighted geometric means are then normalized using their total to obtain the 0-norm probability for an outcome for each outcome as follows:

$$\text{0-norm probability for outcome } j = P_0(j) = \frac{M_0(j)}{\sum_{k=1}^{n} M_0(k)} \quad (5.18)$$

3. **Weighted harmonic average:**
The weighted harmonic mean for outcome j can be computed as follows:

$$\text{Weighted harmonic mean for outcome } j = M_{-1}(j) = \frac{1}{\sum_{i=1}^{E} \frac{w_i}{p_{ji}}} \quad (5.19)$$

The weighted harmonic means are then normalized using their total to obtain the -1-norm probability for an outcome for each outcome as follows:

Chapter five: Consensus and aggregating expert opinions 217

$$\text{–1-norm probability for outcome } j = P_{-1}(j) = \frac{M_{-1}(j)}{\sum_{k=1}^{n} M_{-1}(k)} \quad (5.20)$$

4. **Maximum value:**
The maximum value for outcome j can be computed as follows:

$$\text{Maximum value for outcome } j = M_{\infty}(j) = \max_{i=1}^{E}(p_{ji}) \quad (5.21)$$

The maximum values are then normalized using their total to obtain the ∞-norm probability for an outcome for each outcome as follows:

$$\infty\text{-norm probability for outcme } j = P_{\infty}(j) = \frac{M_{\infty}(j)}{\sum_{k=1}^{n} M_{\infty}(k)} \quad (5.22)$$

5. **Minimum value:**
The minimum value for outcome j can be computed as follows:

$$\text{Minimum value for outcome } j = M_{-\infty}(j) = \min_{i=1}^{E}(p_{ji}) \quad (5.23)$$

The minimum values are then normalized using their total to obtain the $-\infty$-norm probability for an outcome for each outcome as follows:

$$-\infty\text{-norm probability for outcome } j = P_{-\infty}(j) = \frac{M_{-\infty}(j)}{\sum_{k=1}^{n} M_{-\infty}(k)} \quad (5.24)$$

6. **Generalized weighted average:**
The generalized weighted average for outcome j can be computed as follows:

$$\text{Generalized weighted average for outcome } j = M_r(j) = \left(\sum_{i=1}^{E} w_i p_{ji}^r\right)^{1/r} \quad (5.25)$$

The generalized weighted for averages are then normalized using their total to obtain the *r*-norm probability for an outcome each outcome as follows:

$$r\text{-norm probability for outcome } j = P_r(j) = \frac{M_r(j)}{\sum_{k=1}^{n} M_r(k)} \qquad (5.26)$$

where for $r = 1$, and -1, cases 1 and 3 result, respectively.

5.4.4 Uncertainty-based criteria for combining expert opinions

Expert opinions are propositions that do not necessarily meet the JTB requirements of knowledge and, hence, can contain both useful information and uncertainties. In combining these opinions, we have a vested interest in using a process that utilizes all the information contents provided by the experts and that can account for the various uncertainties in producing a combined opinion. The uncertainties can include nonspecificity, conflict, confusion, vagueness, biases, and varying reliability levels of sources, among other types.

Section 5.3 presents methods for measuring uncertainty. These measures deal with the various uncertainty types. In combining expert opinions, we can develop expert-opinion aggregation methods that are either information-based or uncertainty-based. In fact, information and uncertainty can be argued to represent a duality, since information can be considered useful by a cognitive agent if this information results in reducing its uncertainty under prescribed conditions. Therefore, the amount of relevant information gained by the agent is related and can be measured by the amount of uncertainty reduced. This concept of uncertainty-based information was introduced by Klir (1985). The various uncertainty measures presented in this section can deal with various uncertainty types and offer strengths and weaknesses with commensurate complexity and computational demands. The selection of an appropriate uncertainty measure or combinations thereof is problem dependent, and a tradeoff decision between the computational effort needed and the return on this effort in the form of a refined combination of expert opinions needs to be made.

Three uncertainty-based criteria can be used to combine expert opinion: (1) minimum uncertainty, (2) maximum uncertainty, and (3) uncertainty invariance. These three criteria are described in subsequent sections. The criteria of minimum and maximum uncertainty were developed and had great utilities in classical information theory and is commonly referred to the *principles of minimum and maximum entropy*.

5.4.4.1 Minimum uncertainty criterion

Engineers and scientists have a need to select among alternative solutions based on information given on a problem where each solution has a different level of information retention, bias, and error uncertainties. For example, curve fitting of an analytical model to empirical results commonly involves the computation of model parameters with each set of parameters leading to various levels of information retention and uncertainty. An optimal solution in this case can be defined as the solution that maximizes information retention, i.e., minimizes uncertainty. The principle of least squares in regression analysis is an example of such criterion. Other examples include simplification problems in order to deal with systems complexity, and conflict resolution problems in cases of gathering evidence in failure classification.

5.4.4.2 Maximum uncertainty criterion

The criterion of maximum uncertainty can be used in cases involving reasoning that could lead to conclusions that are not necessarily entailed in given premises. In such cases, we should intuitively use all information supported by available evidence but without unintentionally adding information unsupported by the evidence given. This criterion employs the relationship between information and uncertainty by requiring any conclusion resulting from any inference to maximize the relevant uncertainty within constraints representing given premises. As a result, we fully limit our inference ability by our ignorance when making inferences beyond the premise information domain, and we fully utilize information provided by the premises. This criterion, therefore, provides use with assurances of maximizing our nonreliance on information not contained in the premises.

The criterion of maximum uncertainty appeals to engineers and scientists since it results in inferences and solutions that do not go beyond premises given. For example, predictive, scientific models can be viewed as inference models using premises. In system identification, statements on a system or subsystems need to be based on partial knowledge of the system; hence the need to make sure that our inferences do not go beyond information and premises available to us. In selecting a likelihood distribution, the criterion of maximum uncertainty can provide the means of complete uncertainty retention and not assuming additional information beyond what is given to us.

Example 5.1 *Selection of distribution types based on selected constraints using uncertainty measures*

The criterion of maximum uncertainty can be used to select the distribution type that maximizes uncertainty for given constraints. The entropy uncertainty measure was used for this purpose. Table 5.1 summarizes distribution types that maximize uncertainty for a selected list of constraints. For example, the constraints $a \leq X \leq b$ and

Table 5.1 Maximum-Entropy Probability Distributions

Constraints	Maximum-Entropy Distribution
$\int_a^b f_X(x)dx = 1$ Minimum value = a Maximum value = b	Uniform
$\int_0^\infty f_X(x)dx = 1$ Expected value = $\overline{X}$	Exponential
$\int_{-\infty}^\infty f_X(x)dx = 1$ Expected value = $\overline{X}$ Standard deviation = S^2	Normal
$\int_a^b f_X(x)dx = 1$ Expected value = $\overline{X}$ Standard deviation = S^2 Finite range of a minimum value = a, and a maximum value = b	Beta
$\sum_{i=1}^\infty p_i = 1$ where p_i = probability of i independent and identical events occurring in an interval T with an expected rate of occurrence of events of λ	Poisson

$$\int_a^b f_X(x)dx = 1 \qquad (5.27)$$

can be used to maximize the entropy according to Equation 5.7 as follows:

$$\text{maximize } B(f) = -\int_a^b f_X(x)\ln(f_X(x))dx \qquad (5.28)$$

Chapter five: Consensus and aggregating expert opinions 221

Using the method of Lagrange multipliers, the following equation can be obtained:

$$-\frac{\partial}{\partial f}(f\ln(f)) + \lambda\frac{\partial}{\partial f}(f) = 0 \qquad (5.29)$$

This equation has the following solutions:

$$-1 - \ln(f) + \lambda = 0 \qquad (5.30a)$$

$$f_X(x) = e^{\lambda - 1} \qquad (5.30b)$$

Since λ is a constant, f must be a constant leading to the following expression for f:

$$f_X(x) = \frac{1}{b-a} \qquad (5.31)$$

The corresponding entropy is

$$B(f) = \ln(b-a) \qquad (5.32)$$

5.4.4.3 *Uncertainty invariance criterion*

The criterion of uncertainty invariance was developed to facilitate meaningful transformations among various uncertainty measures. This criterion utilizes uncertainty measures that should be carefully constructed in terms of scale and units to allow for transforming one uncertainty type to another; therefore, once all uncertainties are consistently measured, they can be added, manipulated, and treated using a most convenient theory. The criterion of uncertainty invariance principle was used in probability-possibility transformations in combining objective and subjective information that are represented using probability and possibility theory (Brown, 1980).

5.4.5 *Opinion aggregation using interval analysis and fuzzy arithmetic*

Sometimes it might be desirable to elicit probabilities and/or consequences using linguistic terms as shown in Table 4.6 for linguistic probabilities. Linguistic terms of this type can be translated into intervals or fuzzy numbers. Intervals are considered a special case of fuzzy numbers which are in turn a special case of fuzzy sets. Section 4.2.3 shows methods for performing fuzzy arithmetic and fuzzy calculus that can be used to combine expert opinions in this case.

5.4.6 Opinion aggregation using Dempster's rule of combination

Dempster's rule of combination can be used to combine independent expert opinions provided as basic assignments as provided in Section 4.3.3. For two experts that provide the assignments m_1 and m_2 on the same element and a family of sets of interest, the combined pinion ($m_{1,2}$) was given in Equation 4.121 as follows:

$$m_{1,2}(A_i) = \frac{\sum_{\text{all } A_j \cap A_k = A_i} m_1(A_j) m_2(A_k)}{1 - \sum_{\text{all } A_j \cap A_k = \emptyset} m_1(A_j) m_2(A_k)} \quad (5.33)$$

where A_i must be a nonempty set, and $m_{1,2}(\emptyset)=0$. The term $1 - \sum_{\text{all } A_j \cap A_k = \emptyset} m_1(A_j) m_2(A_k)$ of Equation 5.33 is a normalization factor. Equation 5.33 does not account for the reliability of the source and other relevant considerations. Example 4.15 demonstrates the use of the Dempster's rule of combination for classifying a bridge failure case according to the failure causes. Equation 5.33 can result in counterintuitive conclusions in cases of conflicting expert opinions.

5.4.7 Demonstrative examples of aggregating expert opinions

This section contains two example uses of uncertainty measure for aggregating expert opinions (Lai, 1992; Ayyub and Lai, 1992; Lai and Ayyub, 1994). The first example demonstrates the use of uncertainty measure to combine opinions in defining failures; the second example demonstrates the use of uncertainty measures to classify failure to predefined failure categories.

5.4.7.1 Aggregation of expert opinions

The measures of dissonance and confusion, which are constructed in the framework of the theory of evidence, are applied herein for aggregating the expert opinions.

Let $\alpha_1, \alpha_2, \ldots, \alpha_N$ be fuzzy failure definitions for some specified failure mode expressed as structural response and degree of belief for failure definitions that are obtained from experts 1, 2, ..., N, as shown in Figure 5.1. The vertical axis is called failure level; however, it can be viewed as a degree of belief for failure occurrence. These definitions can be viewed as functions representing the same failure state expressed by the N experts. The combined failure definition or function can be obtained by aggregating the N expert opinions as shown in Figure 5.1. The aggregated function is noted as α_0 in Figure 5.1. The lower bound, r_l, and the upper bound, r_u, of structural response for the entire ranges of all functions, and some specified structural response r^* within the lower bound and the upper bound are shown in the figure. In this approach, the values of the N fuzzy failure functions at the

Chapter five: Consensus and aggregating expert opinions

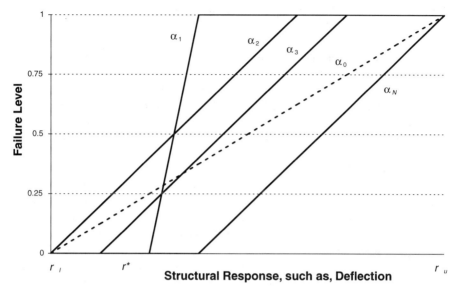

Figure 5.1 Fuzzy failures according to several experts.

specified structural response r^* are interpreted as a basic assignment for experts 1, 2, . . ., N, i.e., $m(\{1\}) = \alpha_1(r^*)$, $m(\{2\}) = \alpha_2(r^*),. . ., m(\{N\}) = \alpha_N(r^*)$. Since each basic assignment is given for the corresponding set of individual experts, there is no evidence supporting the unions of any other combinations of expert opinions. This means that the basic assignment corresponds to the sets of singletons only; however, the summation of all the basic assignments is required to be equal to 1.0. Therefore, if the summation of the basic assignment does not equal 1.0, i.e., $m(\{1\}) + m(\{2\}) + . . . + m(\{N\}) \neq 1.0$, the difference between the summation of the basic assignment and 1 should be distributed to the set of singletons. Since there is no particular preference for any set of individual experts, the difference should be distributed to the sets of singletons by normalization with respect to the summation, such that the normalized summation is equal to 1.0.

Once the basic assignments are properly determined, Equations 5.9 and 5.10 are used, respectively, to calculate the measure of dissonance (D) and the measure of confusion (C) for the specified structural responses. It should be noted that the measure of dissonance is equal to the measure of confusion in this case, since the nonzero basic probability assignments exist only for the sets of singletons. Under this circumstance, both the measures are equal to the Shannon entropy (S) (Shannon, 1948). Therefore, the measure of uncertainty can be calculated as the following:

$$D = C = S = -\sum_{i=1}^{N} m(\{i\})\log_2(i) \qquad (5.34)$$

where $m(\{i\})$ is the adjusted basic assignment for expert i. It is expected that the maximum measure of uncertainty occurs wherever all the experts are of the same opinion at some structural response level, i.e., $\alpha_1(r^*) = \alpha_2(r^*) = \ldots = \alpha_N(r^*) = \alpha_0(r^*)$. Therefore, the closer the experts' opinions to some common level, the larger the measure of uncertainty. The total measure of uncertainty, which is calculated by integrating the measure of uncertainty over the entire range of no common opinion, can be treated as some kind of index to measure the uniformity (or agreement) of the experts' opinions. The closer the experts' opinions to uniformity, the larger the total measure of uncertainty. Therefore, the aggregated linear function α_0 can be obtained by maximizing the total measure of uncertainty.

Considering the resisting-moment versus curvature relationship of the hull structure of a ship subjected to a hogging moment only. The transition from survival to failure in the crisp case was assumed to be attained at curvature level of $\phi_f = 0.3 \times 10^{-5}$. In order to illustrate the application of uncertainty measure in aggregating expert opinions, two fuzzy failure definitions are selected as α_1 and α_2 in Figure 5.2. A linear function α_0 of fuzzy failure definition is considered to be the aggregated expert opinion. The lower bound and the upper bound of curvature range for the fuzzy failure function α_0 are also shown as ϕ_L and ϕ_U in the figure. In this example, the two fuzzy failure definitions are expressed by the following equations:

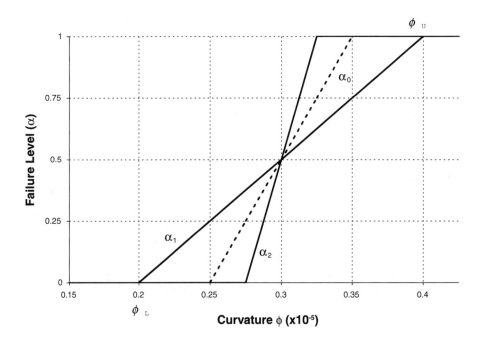

Figure 5.2 Fuzzy failures according to two experts.

Chapter five: Consensus and aggregating expert opinions

$$\text{fuzzy definition 1: } \alpha_1 = 5 \times 10^5 \phi - 1 \tag{5.35a}$$

$$\text{fuzzy definition 2: } \alpha_2 = 20 \times 10^5 \phi - 5.5 \tag{5.35b}$$

The aggregation failure function is assumed in the following linear form:

$$\alpha_0 = a\phi - b \tag{5.36}$$

where a = slope of the linear function of fuzzy failure definition, and b = intercept. The slope a and the intercept b can then be derived as

$$a = \frac{1}{\phi_U - \phi_L} \tag{5.37a}$$

$$b = a\,\phi_L \tag{5.37b}$$

In addition, the aggregated linear function was selected to pass through the point $(\phi, \alpha) = (0.3 \times 10^{-5}, 0.5)$ since the two fuzzy failure functions proposed by experts pass through the same point. Therefore, the parameters ϕ_L and ϕ_U (or a and b) are related. Only one parameter is needed to uniquely define the function α_0. The lower bound ϕ_L of curvature range is chosen as the independent variable that controls the curve α_0. Once the lower bound ϕ_L is assumed, the upper bound ϕ_U can be calculated using the following equation:

$$\phi_U = \phi_L + 2(0.3 \times 10^{-5} - \phi_L) = 0.6 \times 10^{-5} - \phi_L \tag{5.38}$$

The corresponding slope and intercept can then be evaluated using Equations 5.37a and 5.37b. The basic probability assignments for all possible sets of expert opinions are shown in the following:

$$m(\{1\}) = \begin{cases} \alpha_1 + \dfrac{1-(\alpha_1+\alpha_2+\alpha_0)}{3} & \text{if } \alpha_1+\alpha_2+\alpha_0 \le 1 \\[2mm] \dfrac{\alpha_1}{\alpha_1+\alpha_2+\alpha_0} & \text{if } \alpha_1+\alpha_2+\alpha_0 > 1 \end{cases} \tag{5.39a}$$

$$m(\{2\}) = \begin{cases} \alpha_2 + \dfrac{1-(\alpha_1+\alpha_2+\alpha_0)}{3} & \text{if } \alpha_1+\alpha_2+\alpha_0 \le 1 \\[2mm] \dfrac{\alpha_2}{\alpha_1+\alpha_2+\alpha_0} & \text{if } \alpha_1+\alpha_2+\alpha_0 > 1 \end{cases} \tag{5.39b}$$

$$m(\{0\}) = \begin{cases} \alpha_0 + \dfrac{1-(\alpha_1+\alpha_2+\alpha_0)}{3} & \text{if } \alpha_1+\alpha_2+\alpha_0 \le 1 \\ \dfrac{\alpha_0}{\alpha_1+\alpha_2+\alpha_0} & \text{if } \alpha_1+\alpha_2+\alpha_0 > 1 \end{cases} \qquad (5.39c)$$

$$m(\{1,2\}) = m(\{2,0\}) = m(\{0,1\}) = m(\{1,2,0\}) = 0 \qquad (5.39d)$$

Once the basic probability assignments are constructed, the measure of uncertainty can be calculated using Equation 5.34. The results of the total measure of uncertainty for different fuzzy failure definitions are shown in Figure 5.3 and Table 5.2. The aggregated linear function was obtained at the maximum total measure of uncertainty. From the results shown in Figure 5.3, the maximum total measure of uncertainty occurs where the range of curvature is from 0.255 to 0.345 as indicated in Table 5.2. The resulting aggregated fuzzy failure function is, therefore, expressed as

$$\alpha_0 = 11.1 \times 10^5 \phi - 2.83 \qquad (5.40)$$

The slope of this aggregated fuzzy failure function ($a = 11.1 \times 10^5$) is between the two slopes proposed by the experts ($a = 5 \times 10^5$ and 20×10^5 for α_1 and α_2, respectively) and is consistent with engineering intuition.

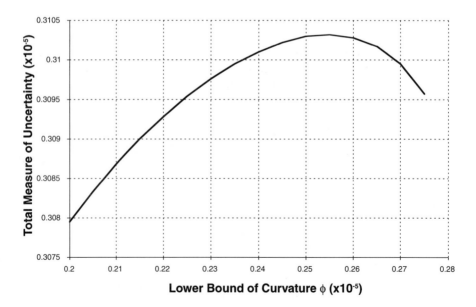

Figure 5.3 Total measure of uncertainty for fuzzy failure definitions.

Chapter five: Consensus and aggregating expert opinions   227

Table 5.2 Total Measure of Uncertainty for Fuzzy Failure Definitions

Lower Bound of Curvature	Slope (a)	Intercept (b)	Total Uncertainty Measure
0.200	5.000	1.000	0.30795
0.205	5.263	1.079	0.30833
0.210	5.556	1.167	0.30868
0.215	5.882	1.265	0.3090
0.220	6.250	1.375	0.30928
0.225	6.667	1.500	0.30954
0.230	7.143	1.643	0.30976
0.235	7.692	1.808	0.30995
0.240	8.333	2.000	0.3101
0.245	9.091	2.227	0.31022
0.250	10.000	2.500	0.3103
0.255	11.111	2.833	Max = 0.31032
0.260	12.500	3.250	0.31028
0.265	14.286	2.786	0.31017
0.270	16.667	4.500	0.30995
0.275	20.000	5.500	0.30957

5.4.7.2   Failure classification

In this case example, six events of structural performance are defined for convenience to track and model structural performance. The following fuzzy events are defined for this purpose as shown in Figure 5.4: complete survival, low serviceability failure, serviceability failure, high serviceability failure, partial collapse, and complete collapse. These events are not necessarily mutually exclusive, as shown in the figure, since intersections might exist between adjacent events. It is of interest to classify some actual (or observed) structural performance or response to one of these failure categories, i.e., events. If the structural response is located within the range of just one event of structural performance, the structural response is classified to this failure category. If the structural response is located over two or more failure events, confusion in classifying the actual structural response into any of the failure categories results. Therefore, the measure of confusion is used in this case for the purpose of failure classification.

Consider an actual structural response ϕ_A which is an observed level that can be represented as event A in Figure 5.5. Categories I and II represent serviceability failure and partial collapse, respectively, according to the expert opinions. Category I is called the lower failure category, and category II is called the higher failure category. Since the magnitude of the structural response ϕ_A is located in the intersection of serviceability failure and partial collapse, confusion exists for the given body of evidence represented by event A and performance categories I and II. Using the measure of confusion, the less distinguishable the two events, the larger degree of confusion between them. Therefore, if event A is less distinguishable with category I rather than category II, event A has a higher level of confusion with category

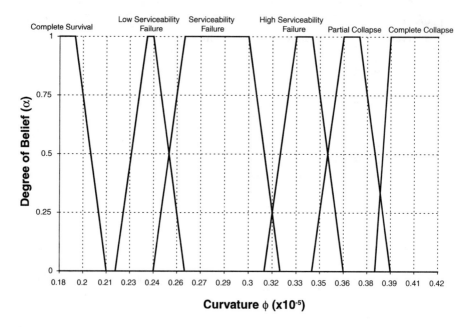

Figure 5.4 Six events of structural response.

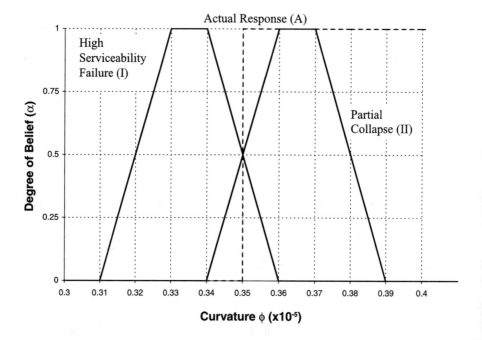

Figure 5.5 Failure events.

Chapter five: Consensus and aggregating expert opinions 229

Table 5.3 The Measure of Confusion for an Actual Response

Parameter	Category I		Category II	
	Actual Response	High Serviceability Failure	Actual Response	Partial Collapse
Basic Assignment (m)	0.639	0.361	0.859	0.141
Belief Measure (Bel)	0.639	0.181	0.930	0.071
Confusion Measure (C)	1.304		0.628	

I than its confusion level with category II. In this case, event A is classified into category I (serviceability failure). One the contrary, if event A has a higher level of confusion with category II, event A is classified into category II (partial collapse). The computations are shown in Table 5.3.

The two events can be used to demonstrate the use of the measure of confusion in failure classification. A specified (or observed) curvature level, $\phi_A = 0.35 \times 10^{-5}$, is first selected as the actual structural response as shown in Figure 5.5. Since this level of damage is located in the intersection of two categories, i.e., high serviceability failure and partial collapse, confusion exists in classifying the observed (specified) structural response to a proper failure category. The measure of confusion is, therefore, computed herein for the purpose of failure classification. The belief measures and the measure of confusion for each body of evidence can also be evaluated and compared. The results are shown in Table 5.3. It is evident from Table 5.3 that the measure of confusion for category I ($C_{A,I} = 1.304$) is larger than the measure of confusion for category II ($C_{A,II} = 0.628$).

Six events of structural performance are assumed in this case as shown in Figure 5.4. These six events were selected for the purpose of illustrating a damage spectrum. The definitions of the six events are not interpreted as "at least"; e.g., although the event serviceability failure is commonly interpreted as at least serviceability failure, it is not treated as such in this example. Therefore, the failure events are treated not nested. Lai (1992) examined both nested and nonnested failure events. In this nonnested case, measures of confusion for the six categories are computed as $\phi_A = 0.35 \times 10^{-5}$ was incrementally increased from the left to the right of Figure 5.4. Since the event with the largest measure of confusion is selected for the structural response classification, the domains of all six classification events can be determined by comparing the degrees of confusions. Event A has a confusion measure with each event gradually increasing until a maximum is reached following a decrease as to moves to the right. Figure 5.6 shows the classification of A based on the confusion measure as a step function. The numbers in the boxes indicate the event classification with numbers 1, 2, 3, 4, 5, and 6 correspond to (1) complete survival, (2) low serviceability failure, (3) serviceability failure, (4) high serviceability failure, (5) partial collapse, and (6) complete collapse, respectively. The confusion measure was computed similarly to the case presented in Table 5.3. It is evident from Figure 5.6 that

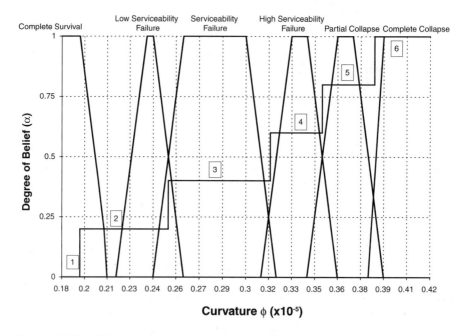

Figure 5.6 Classifying an actual response to the six events of structural response.

the classification of an event changes from a lower failure category to an adjacent higher failure category at a curvature level located near the intersection of the two-adjacent failure categories.

5.5 Exercise problems

Problem 5.1 For the following probability mass function,

x	10	20	30	40	50
$P_X(x)$	0.40	0.30	0.20	0.05	0.05

Compute (a) the Hartley uncertainty measure, and (b) the Shannon entropy uncertainty measure. Discuss their meanings.

Problem 5.2 For the following probability mass function:

x	1	2	3	4	5
$P_X(x)$	0.10	0.20	0.40	0.20	0.10

Compute (a) the Hartley uncertainty measure, and (b) the Shannon entropy uncertainty measure. Discuss their meanings.

Chapter five: Consensus and aggregating expert opinions   231

Problem 5.3 For the triangular set defined as $A = [10, 15, 20]$ and trapezoidal set defined as $B = [10, 12, 18, 20]$, compute their fuzziness uncertainty measures.

Problem 5.4 For the triangular set defined as $A = [10, 15, 20]$ and trapezoidal set defined as $B = [10, 12, 18, 20]$, compute the fuzziness uncertainty measures for outcomes of the following:

1. $A + B$;
2. $A - B$;
3. $A \times B$;
4. A / B;
5. $A + B$, with the constraint that $a = b$ where $a \in A$ and $b \in B$;
6. $A - B$, with the constraint that $a = b$ where $a \in A$ and $b \in B$;
7. $A \times B$, with the constraint that $a = b$ where $a \in A$ and $b \in B$; and
8. A / B, with the constraint that $a = b$ where $a \in A$ and $b \in B$.

Compare and discuss your results.

Problem 5.5 Compute the 25 percentile, median, and 75 percentile aggregation of the following expert opinions:

[0.1, 0.2, 0.1, 0.3, 0.1, 0.2, 0.15]

Problem 5.6 Compute the 25 percentile, median, and 75 percentile aggregation of the following expert opinions:

[100, 120, 110, 100, 90, 150, 110, 120, 105]

Problem 5.7 Five experts provided the following occurrence probabilities for an event E:

[0.001, 0.01, 0.002, 0.008, 0.005]

The experts were assigned the following weight factors based on their abilities as perceived by an analyst:

[0.1, 0.3, 0.25, 0.15, 0.2]

Compute the weighted, aggregated opinion of the five experts using all applicable methods of Section 5.4.3. Compare the results from the various methods and discuss. Provide recommendations on the use of various methods.

Problem 5.8 Describe the method of Lagrange multipliers used in the derivation of Equation 5.32, and rederive Equation 5.32 with details.

Problem 5.9 Two experts classified a bird species to three possible causes for its population decline as provided in the following table in the form of basic assignments m_1 and m_2:

Subset (i.e., cause)	Expert 1 m_1	Expert 2 m_2
Changes in land use (C)	0.05	0.05
Hunting (H)	0.25	0.30
Disease (D)	0.05	0.10
$C \cup H$	0.25	0.05
$C \cup D$	0.05	0.05
$H \cup D$	0.25	0.20
$C \cup H \cup D$	Not provided	Not provided

Compute the belief measures for experts 1 and 2. Compute the basic assignment for the combined judgment and the corresponding belief measure.

Problem 5.10 Three experts classified a bird species to three possible causes for its population decline as provided in the following table in the form of basic assignments m_1, m_2 and m_3:

Subset (i.e., cause)	Expert 1 m_1	Expert 2 m_2	Expert 3 m_3
Changes in land use (C)	0.05	0.05	0.05
Hunting (H)	0.25	0.30	0.50
Disease (D)	0.05	0.10	0.05
$C \cup H$	0.25	0.05	0.01
$C \cup D$	0.05	0.05	0.05
$H \cup D$	0.25	0.20	0.01
$C \cup H \cup D$	Not provided	Not provided	Not provided

Compute the belief measures for experts 1, 2, and 3. Provide procedures for computing the basic assignment for the combined judgment and the corresponding belief measure, and demonstrate the procedures using the above table.

chapter six

Guidance on expert-opinion elicitation

Contents

6.1. Introduction and terminology 233
    6.1.1. Theoretical bases 233
    6.1.2. Terminology 234
6.2. Classification of issues, study levels, experts,
and process outcomes 236
6.3. Process definition 238
6.4. Need identification for expert-opinion elicitation 239
6.5. Selection of study level and study leader 239
6.6. Selection of peer reviewers and experts 241
    6.6.1. Selection of peer reviewers 241
    6.6.2. Identification and selection of experts 241
    6.6.3. Items needed by experts and reviewers before the expert-
opinion elicitation meeting 243
6.7. Identification, selection, and development of technical issues 244
6.8. Elicitation of opinions 245
    6.8.1. Issue familiarization of experts 245
    6.8.2. Training of experts 245
    6.8.3. Elicitation and collection of opinions 246
    6.8.4. Aggregation and presentation of results 247
    6.8.5. Group interaction, discussion, and revision by experts 247
6.9. Documentation and communication 247
6.10. Exercise problems 247

6.1 Introduction and terminology

6.1.1 Theoretical bases

This chapter focuses on occurrence probabilities and consequences of events to demonstrate the process presented in this chapter. For this purpose, the expert-opinion elicitation process can be defined as a formal process of obtaining information or answers to specific questions about certain quantities, called issues, such as failure rates, failure consequences, and expected service life. Expert-opinion elicitation should not be used in lieu of rigorous reliability and risk analytical methods but should be used to supplement them and to prepare for them. The expert-opinion elicitation process presented in this chapter is a variation of the Delphi technique (Helmer, 1968) with scenario analysis (Kahn and Wiener, 1967) based on uncertainty models (Ayyub, 1991, 1992, 1993; Haldar et al., 1997; Ayyub et al., 1997; Ayyub and Gupta, 1997; Ayyub, 1998; Cooke, 1991), social research (Bailey, 1994), USACE studies (Ayyub et al., 1996; Baecher, 1998), ignorance, knowledge, information and uncertainty of Chapter 1, experts and opinions of Chapter 3, nuclear industry recommendations (NRC, 1997), and the Stanford Research Institute protocol (Spetzler and Stael von Holstein, 1975).

6.1.2 Terminology

The terminology of Table 6.1 is needed for defining the expert-opinion elicitation process, in addition to other related definitions provided in Chapters 1, 2, and 3. Table 6.1 provides definitions of terms that related to the expert-opinion elicitation process.

The *expert-opinion elicitation* (EE) process is defined as a formal, heuristic process of gathering information and data or answering questions on issues or problems of concern. The EE process requires the involvement of a leader of EE process who is an entity having managerial and technical responsibility for organizing and executing the project, overseeing all participants, and intellectually owning the results. Experts who render opinions are defined in Chapters 1 and 3. Section 3.2 defines an *expert* as a very skillful person who has much training and knowledge in a special field. The expert is the provider of an opinion in the process of expert-opinion elicitation. An evaluator is an expert who has the role of evaluating the relative credibility and plausibility of multiple hypotheses to explain observations. The process involves evaluators who consider available data, become familiar with the views of proponents and other evaluators, question the technical bases of data, and challenge the views of proponents; it also involves observers who can contribute to the discussion but cannot provide expert opinion that enters in the aggregated opinion of the experts. The process might require peer reviewers who can provide an unbiased assessment and critical review of an expert-opinion elicitation process, its technical issues, and results. Some of the experts might be proponents who are experts who advocate a particular hypothesis or technical position. In science, a proponent evaluates experimental data and professionally offers a hypothesis that would be challenged by the proponent's peers until proven correct or wrong. Resource experts can be used who are technical experts with detailed and deep knowledge of particular data, issue aspects, particular methodologies, or use of

Table 6.1 Terminology and Definitions

Term	Definition
Evaluators	Consider available data, become familiar with the views of proponents and other evaluators, question the technical bases of data, and challenge the views of proponents.
Expert	A person with related or unique experience on an issue or question of interest for the process.
Expert-opinion elicitation (EE) process	A formal, heuristic process of gathering information and data or answering questions on issues or problems of concern.
Leader of EE process	Has managerial and technical responsibility for organizing and executing the project, overseeing all participants, and intellectually owning the results.
Observers	Observers can contribute to the discussion but cannot provide expert opinion that enters in the aggregated opinion of the experts.
Peer reviewers	Experts who can provide an unbiased assessment and critical review of an expert-opinion elicitation process, its technical issues, and results.
Proponents	Experts who advocate a particular hypothesis or technical position. In science, a proponent evaluates experimental data and professionally offers a hypothesis that would be challenged by the proponent's peers until proven correct or wrong.
Resource experts	Technical experts with detailed and deep knowledge of particular data, issue aspects, particular methodologies, or use of evaluators.
Sponsor of EE process	Provides financial support and owns (in the sense of property ownership) the rights to the results of the EE process.
Subject	A person who might be affected by or might affect an issue or question of interest for the process.
Technical facilitator (TF)	Responsible for structuring and facilitating the discussions and interactions of experts in the EE process; stages effective interactions among experts; ensures equity in presented views; elicits formal evaluations from each expert; and creates conditions for direct, noncontroversial integration of expert opinions.
Technical integrator (TI)	Responsible for developing the composite representation of issues based on informed members and/or sources of related technical communities and experts; explains and defends composite results to experts and outside experts, peer reviewers, regulators, and policy makers; and obtains feedback and revising composite results.
Technical integrator and facilitator (TIF)	Responsible for functions of both TI and TF.

evaluators. The sponsor of EE process is an entity that provides financial support and owns the rights to the results of the EE process (ownership is in the sense of property ownership). A subject is a person who might be affected or might affect an issue or question of interest for the process. A technical facilitator (TF) is an entity responsible for structuring and facilitating the discussions and interactions of experts in the EE process; staging effective interactions among experts; ensuring equity in presented views; eliciting formal evaluations from each expert; and creating conditions for direct, noncontroversial integration of expert opinions. A technical integrator (TI) is an entity responsible for developing the composite representation of issues based on informed members and/or sources of related technical communities and experts, for explaining and defending composite results to experts and outside experts, peer reviewers, regulators, and policy makers, and for obtaining feedback and revising composite results. A technical integrator and facilitator (TIF) is an entity responsible for both functions of TI and TF. Table 6.1 provides a summary of these definitions.

6.2 Classification of issues, study levels, experts, and process outcomes

The NRC (1997) classified issues for expert-opinion elicitation purposes into three complexity degrees (A, B, or C), with four levels of study in the expert-opinion elicitation process (I, II, III, and IV) as shown in Table 6.2. A given issue is assigned a complexity degree and a level of study that depend on (1) the significance of the issue to the final goal of the study, (2) the issue's technical complexity and uncertainty level, (3) the amount of nontechnical contention about the issue in the technical community, and (4) important nontechnical considerations such as budgetary, regulatory, scheduling, public perception, or other concerns. Experts can be classified into five types (NRC, 1997): (1) proponents, (2) evaluators, (3) resource experts, (4) observers, and (5) peer reviewers. These terms are defined in Table 6.1.

The study level, as shown in Table 6.2, involves a technical integrator (TI) or a technical integrator and facilitator (TIF). A TI can be one person or a team (i.e., an entity) that is responsible for developing the composite representation of issues based on informed members and/or sources of related technical communities and experts; explaining and defending composite results to experts and outside experts, peer reviewers, regulators, and policy makers; and obtaining feedback and revising composite results. A TIF can be one person or a team (i.e., an entity) that is responsible for the functions of a TI and for structuring and facilitating the discussions and interactions of experts in the EE process; staging effective interactions among experts; ensuring equity in presented views; eliciting formal evaluations from each expert; and creating conditions for direct, noncontroversial integration of expert opinions. The primary difference between the TI and the TIF is in the intellectual responsibility for the study where it lies with only

Table 6.2 Issue Degrees and Study Levels

a. Issue Complexity Degree		b. Study Level	
Degree	Description	Level	Requirements
A	Noncontroversial Insignificant effect on risk	I	TI evaluates and weighs models based on literature review and experience, and estimates needed quantities.
B	Significant uncertainty Significant diversity Controversial Complex	II	TI interacts with proponents and resource experts, assesses interpretations, and estimates needed quantities.
C	Highly contentious Significant effect on risk Highly complex	III	TI brings together proponents and resource experts for debate and interaction. TI focuses the debate, evaluates interpretations, and estimates needed quantities.
		IV	TI and TF (can be one entity, i.e., TIF) organize a panel of experts to interpret and evaluate, focus discussions, keep the experts debate orderly, summarize and integrate opinions, and estimates needed quantities.

Constructed based on NRC, 1997

the TI, and the TIF and experts, respectively. The TIF also has the added responsibility of maintaining the professional integrity of the process and its implementation.

The TI and TIF processes are required to utilize peer reviewers for quality assurance purposes. Two methods of peer review can be performed: (1) participatory peer review, and (2) late-stage peer review. The former method allows for affecting the course of the study, whereas the latter might not be able to affect the study without a substantial rework of the study. The second classification of peer review is peer-review by subject and itself has two types: (1) technical peer review, which focuses on the technical scope, coverage, contents, and results and (2) process peer review, which focuses on the structure, format, and execution of the expert-opinion elicitation

Table 6.3 Guidance on Use of Peer Reviewers

Expert-opinion elicitation Process	Peer Review Subject	Peer Review Method	Recommendation
TIF	Technical	Participatory	Recommended
		Late stage	Can be acceptable
	Process	Participatory	Strongly recommended
		Late stage	Risky; unlikely to be successful
TI	Technical	Participatory	Strongly recommended
		Late stage	Risky but can be acceptable
	Process	Participatory	Strongly recommended
		Late stage	Risky but can be acceptable

NRC, 1997

process. Guidance on the use of peer reviewers is provided in Table 6.3 (NRC, 1997).

The expert-opinion elicitation process should preferably be conducted to include a face-to-face meeting of experts that is developed specifically for the issues under consideration. The meeting of the experts should be conducted after communicating to the experts in advance to the meeting background information, objectives, list of issues, and anticipated outcome from the meeting. The expert-opinion elicitation based on the TIF concept can result in consensus or disagreement as shown in Figure 6.1. Consensus can be of four types as shown in Figure 6.1 (NRC, 1997). Commonly, the expert-opinion elicitation process has the objective of achieving consensus type 4; i.e., experts agree that a particular probability distribution represents the overall scientific community. The TIF plays a major role in building consensus by acting as a facilitator. Disagreement among experts, whether it is intentional or unintentional, requires the TIF to act as an integrator by using equal or nonequal weight factors. Sometimes, expert opinions need to be weighed for appropriateness and relevance rather than strictly weighted by factors in a mathematical aggregation procedure.

6.3 Process definition

Expert-opinion elicitation has been defined as a formal, heuristic process of obtaining information or answers to specific questions about certain quantities, called issues, such as failure rates, failure consequences, and expected service lives. The suggested steps for an expert-opinion elicitation process depend on the use of a TI or a TIF, as shown in Figure 6.2. Figure 6.2 is based on NRC (1997), supplemented with details and added steps. The details of the steps involved in these two processes are defined in subsequent subsections.

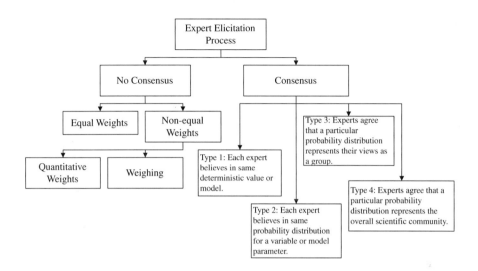

Figure 6.1 Outcomes of the expert-opinion elicitation process.

6.4 Need identification for expert-opinion elicitation

The primary reason for using expert-opinion elicitation is to deal with uncertainty in selected technical issues related to a system of interest. Issues with significant uncertainty, issues that are controversial and/or contentious, issues that are complex, and/or issues that can have a significant effect on risk are most suited for expert-opinion elicitation. The value of the expert-opinion elicitation comes from its initial intended uses as a heuristic, not a scientific, tool, for exploring vague and unknown issues that are otherwise inaccessible. As stated previously, it is not a substitute for scientific, rigorous research.

The identification of need and its communication to experts are essential for the success of the expert-opinion elicitation process. The need identification and communication should include the definition of the study's goal and relevance of issues to this goal. Establishing this relevance makes the experts stake holders and thereby increases their attention and sincerity levels. Relevance of each issue and/or question to the study needs to be established. This question-to-study relevance is essential to enhancing the reliability of collected data from the experts. Each question or issue needs to be relevant to each expert, especially when dealing with subjects with diverse views and backgrounds.

6.5 Selection of study level and study leader

The goal of a study and nature of issues determine the study level as shown in Table 6.2. The study leader can either be a technical integrator (TI),

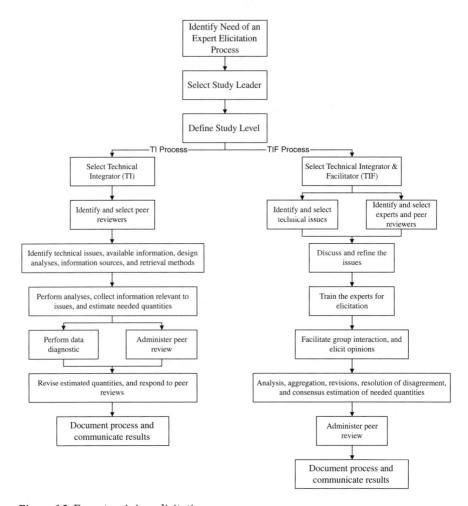

Figure 6.2 Expert-opinion elicitation process.

technical facilitator (TF), or a combined technical integrator and facilitator (TIF). The leader of the study is an entity having managerial and technical responsibility for organizing and executing the project, overseeing all participants, and intellectually owning the results. A study leader should be selected based on the following attributes:

1. An outstanding professional reputation and wide recognition and competence based on academic training and relevant experience;
2. Strong communication skills, interpersonal skills, flexibility, impartiality, and ability to generalize and simplify;
3. A large contact base of industry leaders, researchers, engineers, scientists, and decision makers; and
4. Leadership qualities and the ability to build consensus.

The study leader does not need to be a subject expert, but should be knowledgeable of the subject matter.

6.6 Selection of peer reviewers and experts

6.6.1 Selection of peer reviewers

Peer review can be classified according to the peer-review method and according to the peer-review subject. There are two methods of peer review: (1) participatory peer review which is conducted as an ongoing review throughout all study stages, and (2) late-stage peer review which is performed as the final stage of the study. Peer-review subject also has two types: (1) technical peer review and (2) process peer review. These classifications were discussed in Section 6.2.

Peer reviewers are needed for both the TI and TIF processes and should be selected by the study leader in close consultation with perhaps the study sponsor. Researchers, engineers, and scientists with the following characteristics should be sought as peer reviewers:

- Outstanding professional reputation, and widely recognized competence based on academic training and relevant experience,
- General understanding of the issues in other related areas and/or with relevant expertise and experiences from other areas,
- Available and willing to devote the needed time and effort, and
- Strong communication skills, interpersonal skills, flexibility, impartiality, and ability to generalize and simplify.

6.6.2 Identification and selection of experts

The size of an expert panel should be determined on a case-by-case basis. The panel should be large enough to achieve a needed diversity of opinion, credibility, and result reliability. In recent expert-opinion elicitation studies, a nomination process was used to establish a list of candidate experts by consulting archival literature, technical societies, governmental organizations, and other knowledgeable experts (Trauth et al., 1993). Formal nomination and selection processes should establish appropriate criteria for nomination, selection, and removal of experts. For example, these criteria were used in an ongoing Yucca Mountain seismic hazard analysis (NRC, 1997) to select experts:

1. Strong relevant expertise through academic training, professional accomplishment and experiences, and peer-reviewed publications;
2. Familiarity and knowledge of various aspects related to the issues of interest;
3. Willingness to act as proponents or impartial evaluators;

4. Availability and willingness to commit needed time and effort;
5. Specific related knowledge and expertise of the issues of interest;
6. Willingness to effectively participate in needed debates, to prepare for discussions, and provide needed evaluations and interpretations; and
7. Strong communication skills, interpersonal skills, flexibility, impartiality, and ability to generalize and simplify.

In this NRC study, criteria for removing experts included failure to perform according to commitments and demands as set in the selection criteria, and unwillingness to interact with members of the study.

The panel of experts for an expert-opinion elicitation process should have a balance and broad spectrum of viewpoints, expertise, technical points of view, and organizational representation. The diversity and completeness of the panel of experts is essential for the success of the elicitation process. For example, it can include the following:

1. Proponents who advocate a particular hypothesis or technical position;
2. Evaluators who consider available data, become familiar with the views of proponents and other evaluators, question the technical bases of data, and challenge the views of proponents; and
3. Resource experts who are technical experts with detailed and deep knowledge of particular data, issue aspects, particular methodologies, or use of evaluators.

The experts should be familiar with the design, construction, operational, inspection, maintenance, reliability, and engineering aspects of the equipment and components of a facility of interest. For this study, it was essential to select people with basic engineering or technological knowledge; however they do not necessarily need to be engineers. It might be necessary to include one or two experts from management with engineering knowledge of the equipment and components, consequences, safety aspects, administrative and logistic aspects of operation, expert-opinion elicitation process, and objectives of this study. One or two experts with a broader knowledge of the equipment and components might be needed. Also, one or two experts with a background in risk analysis and risk-based decision making and their uses in areas related to the facility of interest might be needed.

Observers can be invited to participate in the elicitation process and can contribute to the discussion, but they cannot provide expert opinion that enters in the aggregated opinion of the experts. The observers provide expertise in the elicitation process, probabilistic and statistical analyses, risk analysis, and other support areas. The composition and contribution of the observers are essential for the success of this process. The observers may include the following:

- Individuals with research or administrative-related background from research laboratories or headquarters of the U.S. Army Corps of Engineers with engineering knowledge of equipment and components of Corps facilities.
- Individuals with expertise in probabilistic analysis, probabilistic computations, consequence computations and assessment, and expert-opinion elicitation.

A list of names with biographical statements of the study leader, technical integrator, technical facilitator, experts, observers, and peer reviewers should be developed. All attendees can participate in the discussions during the meeting. However, only the experts can provide the needed answers to questions on the selected issues. The integrators and facilitators are responsible for conducting the expert-opinion elicitation process. They can be considered to be a part of the observers or experts depending on the circumstances and the needs of the process.

6.6.3 Items needed by experts and reviewers before the expert-opinion elicitation meeting

The experts and observers need to receive the following items before the expert-opinion elicitation meeting:

- An objective statement of the study;
- A list of experts, observers, integrators, facilitators, study leader, sponsors, and their biographical statements;
- A description of the facility, systems, equipment, and components;
- Basic terminology and definitions that may include probability, failure rate, average time between unsatisfactory performances, mean (or average) value, median value, and uncertainty;
- Failure consequence estimation;
- A description of the expert-opinion elicitation process;
- A related example of the expert-opinion elicitation process and its results, if available;
- Aggregation methods of expert opinions, such as computations of percentiles;
- A description of the issues in the form of a list of questions with background descriptions. Each issue should be presented on a separate page with spaces for recording an expert's judgment, any revisions and comments. Clear statements of expectations from the experts in terms of time, effort, responses, communication, and discussion style and format.

It might be necessary to personally contact individual experts for the purpose of establishing clear understanding of expectations.

6.7 Identification, selection and development of technical issues

The technical issues of interest should be carefully selected to achieve certain objectives. In these guidelines, the technical issues are related to the quantitative assessment of failure probabilities and consequences for selected components, subsystems, and systems within a facility. The issues should be selected such that they would have a significant impact on the study results. These issues should be structured in a logical sequence, starting with background statement, followed by questions, and then answer selections or answer format and scales. Personnel with risk-analysis background who are familiar with the construction, design, operation, and maintenance of the facility need to define these issues in the form of specific questions. Also, background materials about these issues need to be assembled. The materials will be used to familiarize and train the experts about the issues of interest as described subsequent steps.

An introductory statement for the expert-opinion elicitation process should be developed that includes the goal of the study and establishes relevance. Instructions should be provided with guidance on expectations, answering questions, and reporting. The following are guidelines on constructing questions and issues based on social research practices (Bailey, 1994):

1. Each issue can include several questions; however, each question should elicit only one opinion. It is a poor practice to include two questions in one.
2. Question and issue statements should not be ambiguous. Also, the use of ambiguous words should be avoided. In expert-opinion elicitation of failure probabilities, the word "failure" might be vague or ambiguous to some subjects. Special attention should be given to its definition within the context of each issue or question. The level of wording should be kept to a minimum. Also, word choice might affect the connotation of an issue.
3. The use of factual questions is preferred over abstract questions. Questions that refer to concrete and specific matters result in concrete and specific answers which is what is wanted.
4. Questions should be carefully structured in order to reduce biases of subjects. Questions should be asked in a neutral format; sometimes it is more appropriate not to have lead statements.
5. Sensitive topics might require stating questions with lead statements to establish supposedly accepted social norms in order to encourage subjects to answers the questions truthfully.

Questions can be classified into open-ended questions and closed-ended questions as previously discussed. The format of the question should be selected carefully. The format, scale, and units for the response categories should be selected to best achieve the goal of the study. The question order and minimum number of questions should be selected, using the guidelines of Section 3.8.

Once the issues are developed, they should be pretested by administering them to a few subjects for the purpose of identifying and correcting flaws. The results of this pretesting should be used to revise the issues.

6.8 Elicitation of opinions

The elicitation of opinions process should be systematic for all the issues, according to the steps presented in this section.

6.8.1 Issue familiarization of experts

The background materials that were assembled in the previous step should be sent to the experts one to two weeks in advance of the meeting to provide sufficient time for them to become familiar with the issues. The objective of this step is also to ensure that there is a common understanding among the experts of the issues. The background material should include the objectives of the study, description of the issues, lists of questions for the issues, description of systems and processes, their equipment and components, the description of elicitation process and of the method for selecting experts, and biographical information on the selected experts. Also, example results and their meaning, methods of analysis of the results, and lessons learned from previous elicitation processes should be made available. It is important to break the questions or issues into components that can be easily addressed. Preliminary discussion meetings or telephone conversations between the facilitator and experts might be necessary in some cases in preparation for the elicitation process.

6.8.2 Training of experts

Expert training is performed during the meeting of the experts, observers, and facilitators. During the training, the facilitator needs to remain flexible to refining wording or even changing the approach based on feedback from experts. For instance, experts might not be comfortable with the term "probability" but may answer about "events per year" or "recurrence interval." This indirect elicitation, as discussed in Section 3.6.1, should be explored with the experts.

The meeting should be started with presentations of background material to establish relevance of the study to the experts and study goals in order to establish rapport with the experts. Then, information on uncertainty sources and types, occurrence probabilities and consequences, the expert-opinion elicitation process, technical issues and questions, and aggregation of expert opinions should be presented. Also, experts need to be trained on providing answers in an acceptable format that can be used in the analytical evaluation of the failure probabilities or consequences and in areas such as the meaning of "probability," "central tendency," and "dispersion measures," especially those experts who are not familiar with the language of probability. Additional training might be needed on consequences, subjective assessment, logic trees, problem-structuring tools (such as influence diagrams), and methods of combining expert opinions.

Sources of bias (including overconfidence and base-rate fallacy) and their contribution to bias and error should be discussed. This step should include a search for any motivational bias of experts, for example, that due to previous positions taken in public, desire to influence decisions and funding allocations, preconceived notions that experts will be evaluated by their superiors as a result of their answers, and/or to be perceived as an authoritative expert. These motivational biases, once identified, can sometimes be overcome by redefining the incentive structure for the experts.

6.8.3 Elicitation and collection of opinions

The opinion elicitation step starts with a technical presentation of an issue and by decomposing the issue to its components, discussing potential influences and describing event sequences that might lead to top events of interest. These top events are the basis for questions related to the issue in the next stage of the opinion elicitation step. Factors, limitations, test results, analytical models, and uncertainty types and sources need to be presented. The presentation should allow for questions to eliminate any ambiguity and to clarify scope and conditions for the issue. Discussion of the issue should be encouraged and might result in refining the definition of the issue. Then, a form with a statement of the issue should be given to the experts to record their evaluation or input. The experts' judgment along with their supportive reasoning should be documented about the issue.

It is common to ask experts to provide several conditional probabilities in order to reduce the complexity of the questions and thereby obtain reliable answers. These conditional probabilities can be based on fault tree and event tree diagrams. Conditioning has the benefit of simplifying the questions by decomposing the problems. It also results in a conditional event that has a larger occurrence probability than its underlying events; thereby making the elicitation less prone to biases since experts tend to have a better grasp of larger probabilities in comparison to very small ones. It is desirable to have the elicited probabilities in the range of 0.1 to 0.9 if possible. Sometimes it might be desirable to elicit conditional probabilities using linguistic terms, as shown in Table 4.6. If correlation among variables exits, it should be presented to the experts in great detail and conditional probabilities need to be elicited.

Issues should be considered individually, one at a time, although sometimes similar or related issues might be considered simultaneously.

6.8.4 Aggregation and presentation of results

The collected assessments from the experts for an issue should be assessed for internal consistency, analyzed, and aggregated to obtain composite judgments for the issue. The means, medians, percentile values, and standard deviations need to be computed for the issues. Also, a summary of the reasoning provided during the meeting about the issues needs to be developed. Uncertainty levels in the assessments should also be quantified. Methods for combining expert opinions are provided in Chapter 5. The methods can be classified into consensus methods and mathematical methods. The mathematical methods can be based on assigning equal weights or different weights to the experts.

6.8.5 Group interaction, discussion, and revision by experts

The aggregated results need to be presented to the experts for a second round of discussion and revision. The experts should be given the opportunity to revise their assessments of the individual issues at the end of discussion. Also, they should be asked to state the rationale for their statements and revisions. The revised assessments of the experts need to be collected for aggregation and analysis. This step can produce either consensus or no consensus as shown in Figure 6.1. The selected aggregation procedure might require eliciting weight factors from the experts. In this step the technical facilitator plays a major role in developing a consensus, and maintaining the integrity and credibility of the elicitation process. Also, the technical integrator is needed to aggregate the results without biases with reliability measures. The integrator might need to deal with varying expertise levels for the experts, outliers (i.e., extreme views), nonindependent experts, and expert biases.

6.9 Documentation and communication

A comprehensive documentation of the process is essential in order to ensure acceptance and credibility of the results. The document should include complete descriptions of the steps, the initial results, revised results, consensus results, and aggregated results spreads and reliability measures.

6.10 Exercise problems

Problem 6.1 What are the differences between the technical facilitator, and technical integrator, and facilitator in an expert-opinion elicitation process?

Problem 6.2 What are the success requirements for selecting experts and developing an expert panel? How many experts would you recommend? For your range on the number of experts, provide guidance in using the lower and upper ends of the range.

Problem 6.3 Working in teams, select five classmates as a panel of experts and elicit their opinions on five forecasting issues in engineering. Select these issues such that the classmates can pass the test of experts on these issues. Perform all the steps of expert-opinion elicitation; document your process and results as a part of solving this problem.

chapter seven

Applications of expert-opinion elicitation

Contents

7.1. Introduction 249
7.2. Assessment of occurrence probabilities 250
    7.2.1. Cargo elevators onboard ships 250
        7.2.1.1. Background..250
        7.2.1.2. Example issues and results 250
    7.2.2. Navigation locks...251
        7.2.2.1. Background..251
        7.2.2.2. General description of lock operations...........................251
        7.2.2.3. Description of components...255
        7.2.2.4. Example issues and results 257
7.3. Economic consequences of floods..262
    7.3.1. Background 262
    7.3.2. The Feather River basin 262
        7.3.2.1. Levee failure and consequent flooding...........................262
        7.3.2.2. Flood characteristics..262
        7.3.2.3. Building characteristics..263
        7.3.2.4. Vehicle characteristics 264
    7.3.3. Example issues and results...264
        7.3.3.1. Structural depth-damage relationships 264
        7.3.3.2. Content depth-damage relationships..............................264
        7.3.3.3. Content-to-structure value ratios.....................................266
        7.3.3.4. Vehicle depth-damage relationship 266

7.1 Introduction

This chapter demonstrates the application of expert-opinion elicitation by focusing on occurrence probabilities and consequences of events related to naval and civil works systems for the purposes of planners, engineers, and others, should they choose to use expert judgment. For this purpose, formal processes of obtaining information or answers to specific questions about certain quantities, called issues, such as failure rates and probabilities, and failure consequences were conducted, and opinions were elicited and analyzed as presented in this chapter. The guidance on expert-opinion elicitation of Chapter 6 was used for this purpose. The chapter is divided into sections that deal with using expert-opinion elicitation in assessing failure probabilities and failure consequences.

7.2 Assessment of occurrence probabilities

7.2.1 Cargo elevators onboard ships

7.2.1.1 Background

This example illustrates the use of expert-opinion elicitation to obtain failure probabilities needed to study the safety of cargo elevators onboard naval ships (Ayyub, 1992). In order to study the safety of the elevators and the effect of add-on safety features, a fault tree analysis was performed. The fault tree analysis requires the knowledge of failure probabilities of basic events, such as the unsatisfactory performance of mechanical or electrical components and human errors.

Generally, the failure probabilities can be obtained from several sources, such as failure records, failure databases, literature review, or industry-based reports and documents. However, in some cases these sources do not contain the needed probabilities for some basic events. In such cases, expert-opinion elicitation can be used to obtain the needed information. For example, the failure rate of the hoisting machinery brake was obtained from failure records, and the probability that a passerby would fall into an open elevator trunk (human error) required expert-opinion elicitation.

In the elevator safety study, about 250 issues were identified for the expert-opinion elicitation process. The issues were presented to the experts with the needed background information over a three-day period. All the issues were discussed and addressed in this time period.

This section provides example issues and results of expert-opinion elicitation. Since the background information on the types of elevators and their use and limitation are not provided in this section, the results reported herein can be considered to be hypothetical and should not be used for other purposes.

7.2.1.2 Example issues and results

Two example issues are described in this section:

1. How often does the load on a platform shift as a result of being poorly stacked?

2. During one loading revolution at one deck level, what is the probability that a fork truck driver will place the load such that it overhangs the edge of the platform?

Eight experts were used in the expert-opinion elicitation process. The results of the process were summarized in the form of percentiles. The percentiles were computed using the equations in Table 4.7. Tables 7.1 and 7.2 summarize the results of the expert-opinion elicitation for issues 1 and 2, respectively. The results are expressed as the number of unsatisfactory performances per year and a percent for issues 1 and 2, respectively. These results were used to compute the needed probabilities in the fault tree analysis. It is desirable in expert-opinion elicitation to state the issues in the most suitable form and units in order to obtain the best results from the experts.

7.2.2 Navigation locks

7.2.2.1 Background

Detailed descriptions of technical issues are essential for the success of an expert-opinion elicitation process and need to be provided to the experts. The descriptions should provide the experts of background materials, clear statements of issues, objectives, formats, and opinion aggregation that would be used in elicitation sessions. In this example, a description of a navigation lock and fault scenarios are presented for demonstration purposes. The equipment and components are based on the Emsworth navigation lock on the Ohio River. Ayyub et al. (1996) used technical background materials on the operations of the lock to develop relevant technical issues.

A navigation lock can be considered to constitute a system that consists of equipment which consists of components consisting of elements. The equipment, components, and elements are called the levels of analysis. In estimating failure likelihood and consequences, decisions need to be made on the level of computations for an item in the process, i.e., equipment, component, or element level. The decision can be based on the availability of information, the logistic of inspection that might define the entity or unit, the objectives of risk analyses performed on the lock, or other considerations. Accordingly, the level of computations does not need to be the same for all items within the process.

7.2.2.2 General description of lock operations

The operation of the lock is shown in the logic diagram in Figures 7.1a and 7.2b (Ayyub et al., 1996).

The lock consists of two adjacent, parallel lock chambers located along the right bank of the main channel. The large lock chamber occupies the landward position and has clear dimensions of 110 ft × 600 ft. The smaller river chamber measures 56 ft × 360 ft. Normal lift is 18 feet. The lock walls and sills are the gravity type and founded on rock. Both the upper and lower

Table 7.1 Expert-Opinion Elicitation for Example Issue 1

Event Name	Full Description	Expert-opinion elicitation (8 experts)					Summary
		First Response	Median	Second Response	Median		
Load is poorly stacked.	The load on the platform is stacked in such a manner that it is shifted by normal starting and stopping of the platform. Assume that the ship is in calm sea state.	Issue: 1 in 1 yr 1 in 1 yr 1 in 0.5 yr 1 in 2 yrs 1 in 0.1 yr	1 in 1 yr	Issue: 1 in 1 yr 1 in 1 yr 1 in 0.5 yr 1 in 1 yr 1 in 0.5 yr	1 in 1 yr		
	Issue: On one elevator, how often does the load on the platform shift as a result of being poorly stacked?	1 in 1 yr 1 in 0.1 yr 1 in 15 yr		1 in 1 yr 1 in 0.5 yr 1 in 1 yr			Low 1 in 1 year 25 percentile 1 in 1 year Median 1 in 1 year 75 percentile 1 in 0.5 year High 1 in 0.5 year

Ayyub, 1992; Ayyub et al., 1996

Chapter seven: Applications of expert-opinion elicitation 253

Table 7.2 Expert-Opinion Elicitation for Example Issue 2

Event Name	Full Description	Expert-opinion elicitation (8 experts)				Summary
		First Response	Median	Second Response	Median	
Fork truck driver places load over- hanging platform.	Fork truck driver places load such that it overhangs platform despite the existence of adequate lighting. Assume that there are no yellow margins painted on the platform.	Issue: 1% 1% 10% 0.1% 0.5% 1% 0.5% 0.5%	0.75%	Issue: 1% 1% 10% 1% 0.5% 1% 0.5% 0.5%	1%	Low 0.5% 25 percentile 0.5% Median 1% 75 percentile 1% High 10%
	Issue: During one loading evolution at one deck level, what is the probability that a fork truck driver will place the load such that it overhangs the edge of the platform?					

Ayyub, 1992; Ayyub et al., 1996

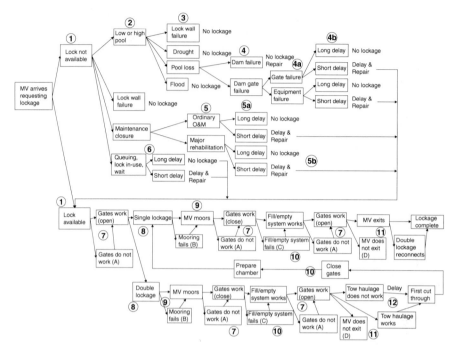

Figure 7.1a Emsworth Navigation Lock on the Ohio River (Ayyub et al., 1996).

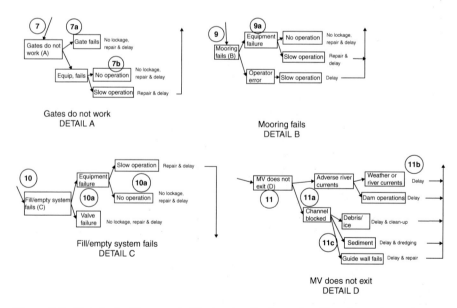

Figure 7.1b Details for Emsworth Navigation Lock on the Ohio River (Ayyub et al., 1996).

guide and guard walls are concrete gravity sections, but the upper and lower guard walls have been extended using steel sheet pile cells. The filling and emptying of the lock chambers is accomplished through ports in the middle and river walls. The large chamber is filled by 16 cylindrical valves located in the upper end of the middle wall and emptied by 16 similar valves which pass the water through a culvert under the smaller chamber and into the river below the dam. A supplemental filling system was instituted during a recent major rehabilitation and involved the reopening of a 10-ft diameter turbine tunnel, providing a slide gate, plugging the tailrace exit, and cutting filling ports through the land wall at lock floor level. The small chamber uses only six filling and six emptying valves in the river wall. The lock gates are of the mitering type, hinged to embedded anchorages at the top and supported at the bottom on steel pintles. Each leaf is a rectangular frame with vertical girders at each end, and vertical beams and horizontal intercostals on the gate leaves for the 110-ft chamber, or horizontal beams and vertical intercostals on the leaves for the 56-ft chamber. Upstream closure of the large chamber is accomplished using trestles stored underwater that are raised from notches in a concrete sill upstream of the miter gates and then fitted with bulkheads. The small chamber uses a coffer beam and needle type closure. Downstream closure for both chambers is accomplished with poiree dams. The average number of annual lockages has remained fairly constant over the last 30 years at about 9950, with commercial lockages decreasing and recreational lockages increasing in recent years.

7.2.2.3 Description of components

As a system, the Emsworth navigation lock on the Ohio River consists of gates, dam, walls, channel, equipment, and users. The following are descriptions of its components:

1. **Filling and emptying valves:** The filling and emptying of the lock chambers are accomplished through culverts placed in the middle and river walls. The main lock is filled by 16 cylindrical valves located in the upper end of the middle wall and emptied by 16 similar valves which pass the water through the lower end of the wall and under the riverward chamber into the river below the dam.
2. **Filling and emptying equipment:** The hydraulic system consists of three constant delivery oil pumps and one pressure-holding oil pump, located on the first floor in the operation building on the land wall. The pumps supply oil under pressure to the hydraulic piping system for operation of the lock gate and culvert valve operating machinery on the lock walls. This system was installed in 1968 and replaced the original compressed air system for operation of the miter gates and the original hydraulic system installed for operation of the emptying and filling valves.

3. **Lock wall:** The lock walls are the gravity type founded on rock. Width of the wall at the top is 5 ft minimum and 24 ft maximum. The sills are concrete gravity sections and anchor rods are installed where computations indicated their need.
4. **Guide wall:** The upper guide wall is 1023.19 ft long measured from the upstream nose of the middle wall, and the lower guide wall is 650.0 ft long measured from the downstream nose of the middle wall. They are gravity structures founded on rock, except for the upper guide wall extension which is constructed of individual steel sheet pile cells.
5. **Miter gates:** The lock gates are constructed of structural steel shapes and plates. The gate leaves for the 110-ft chamber are vertically framed. Each gate consists of two leaves that are hinged to embedded anchorages at the top by gudgeon pins and are supported at the bottom on steel pintles, with the pintle bases embedded in concrete. Each leaf is a rectangular frame with vertical quoin and miter girders at the fixed and free ends, respectively, and vertical beams and horizontal intercostals on the gate leaves for the 110-ft chamber.
6. **Miter gate-operating equipment:** The hydraulic system consists of three constant delivery oil pumps and one pressure holding oil pump, located on the first floor in the operation building on the land wall. The pumps supply oil under pressure to the hydraulic piping system for operation of the lock gate and culvert valve operating machinery on the lock walls. This system was installed in 1968 and replaced the original compressed air system for operation of the miter gates and the original hydraulic system installed for operation of the emptying and filling valves.
7. **Dam gates:** The 13 submergible lift gates are steel structures arranged to travel on vertical tracks on the piers. Each gate can be raised to a point where its bottom is 39.4 ft above the sill and lowered to a point where its top is 3 ft below normal pool level. There is one Sidney gate located on the back channel dam. This gate combines features of both the tainter and vertical lift gates. The gate works like a tainter gate until the gate reaches the limits of its rotation, after which the entire gate is raised by the lifting chains up to the maximum travel limit, 38 ft above the sill.
8. **Dam gate-operating equipment:** Two hoist motors and two synchronous tie motors of the slip-ring induction type are provided for each gate. A full magnetic reverse control panel operates the two hoist motors and the two synchronous tie motors for each gate from a remotely mounted master switch. In the case of emergency, either hoisting motor may be cut out by means of switches, and the gate can be operated by the remaining motor through the synchronous tie motors.

9. **Tow haulage unit:** All the tow haulage equipment is located on the middle wall and is used to assist tows in leaving the 110-ft land chamber. This equipment consists of the following: an electric motor-driven pump; hydraulic motor-driven grooved winch drum; towing bitt; controls; and miscellaneous items, including rails, wire rope, and sheaves. The system is designed for towing a maximum load of 18,000 pounds at a speed of 70 ft-per-min.
10. **Mooring equipment:** There are 20 check posts present for the 110-ft land chamber, 10 on the land wall and 10 on the land side of the middle wall. These are embedded on the top of the walls for routine tow stopping. One floating mooring bitt was installed on the land wall of the 110-ft chamber during the major rehabilitation in 1982. This installation facilitates locking through up-bound tows.

7.2.2.4 Example issues and results

Four example issues are described in this section, two on occurrence likelihood and two on occurrence consequences:

1. The lock is either available when the tow arrives and requests lockage or it is not. What fraction of the total number of tows find the chamber available upon arrival? The answer should be provided in the form of a fraction of the total number of tow arrivals.
2. The chamber has been raised or lowered and the gates are open. The tow is ready to exit the chamber; however, it is unable to exit the chamber due to channel blockage. Please answer the question considering all incidents where the channel is blocked. Given that the tow does not exit the chamber due to channel blockage, what fraction of these occurrences are caused by debris or ice?
3. The lock is not available when the tow arrives and requests lockage due to low pool caused by pool loss. Given that the pool loss is caused by the failure of a dam gate or operating equipment, what are the consequences, repair cost only in $1000, associated with the pool loss caused by failure of the gate itself?
4. The lock is not available when the tow arrives and requests lockage due to closure for repairs or maintenance. Given that the chamber is closed for maintenance or repairs due to ordinary maintenance, how many hours would a typical tow be delayed?

Thirteen experts were used in the expert-opinion elicitation process. The results of the process were summarized in the form of minimum, maximum, and median values. The median was computed using the equations in Table 4.7. Tables 7.3, 7.4, 7.5, and 7.6 summarize the results of the expert-opinion elicitation for issues 1, 2, 3, and 4, respectively. These results were then used in assessing the logic diagrams of Figures 7.1a and 7.1b.

Table 7.3 Expert-Opinion Elicitation for Example Lock Issue 1

Project: Emsworth Lock and Dam Ohio River, Pittsburgh, PA	Description of Component and Issues	Expert-opinion elicitation (8 experts)			
		First Response	Median	Second Response	Median
Chamber Availability	110' Chamber				
	Condition: The lock is either available when the tow arrives and requests lockage or it is not.				
	Please answer the following questions. Frame your response in the form of a fraction of the total number of tow arrivals:				
	1. What fraction of the total number of tows find the chamber available upon arrival?*	0.5		0.5	
		0.95		0.95	
		0.75		0.75	
		0.7		0.7	
	2. What fraction of the total number of tows find the chamber not available upon arrival?	0.38		0.38	
		0.5		0.5	
		0.6		0.6	
		0.75		0.75	
	NOTE: Your answers to questions 1 and 2 above must add up to 1.	0.65		0.65	
		0.3		0.3	
		0.85		0.85	
		0.25	0.25	0.25	0.25	Low
		0.8	0.65	0.8	0.65	Median
			0.95		0.95	High

*Only answers to question 1 are provided herein. Ayyub et al., 1996.

Chapter seven: Applications of expert-opinion elicitation 259

Table 7.4 Expert-Opinion Elicitation for Example Issue 2

Project: Emsworth Lock and Dam, Ohio River, Pittsburgh, PA	Description of Component and Issues	Expert-opinion elicitation (8 experts)			
		First Response	Median	Second Response	Median
	110' Chamber				
The chamber has been raised or lowered and the gates are open.	Condition: The tow is ready to exit the chamber; however, it is unable to exit the chamber due to channel blockage. Please answer the following questions considering all incidents where the channel is blocked: Given that the tow does not exit the chamber due to channel blockage: 1. What fraction of these occurences are caused by debris or ice?*	0.9 0.98 0.98 0.95 0.6		0.9 0.98 0.98 0.95 0.6	
	2. What fraction of these occurrences are caused by sediment buildup?	0.4 0.99 0.9 0.95 0.98		0.4 0.99 0.9 0.95 0.98	
	3. What fraction of these occurrences are caused by a failed guide wall?	0.968 0.91 0.99	0.4 0.95 0.99	0.968 0.91 0.99	0.4 <u>Low</u> 0.95 <u>Median</u> 0.99 <u>High</u>
	Note: Your responses to questions 1, 2, and 3 above must add up to 1.				

Only answers to question 1 are provided herein. Ayyub et al., 1996.

Table 7.5 Expert-Opinion Elicitation for Example Lock Issue 3

Project: Emsworth Lock and Dam, Ohio River, Pittsburgh, PA	Description of Condition and Issues	Expert Elicitation			
		First Response	Median	Second Response	Median
Chamber availability: Not available due to low or high Pool. Consequences	110' Chamber Condition: The lock is not available when the tow arrives and requests lockage due to low pool caused by pool loss. Given that the pool loss is caused by the failure of a dam gate or operating equipment: Please answer the following questions considering all tows which arrive to find this condition: What are the consequences associated with the pool loss caused by failure of the gate itself: Repair costs? (in $1,000)	$100 $100 $500 $50 $81 $100 $25 $25 $2 $10 $50 $250 $250	$2 $81 $500	$100 $100 $500 $50 $81 $100 $25 $25 $2 $10 $50 $250 $250	$2 Low $81 Median $500 High

Chapter seven: Applications of expert-opinion elicitation 261

Table 7.6 Expert-Opinion Elicitation for Example Lock Issue 4

Project: Emsworth Lock and Dam Ohio River, Pittsburgh, PA	Description of Condition and Issues 110' Chamber	Expert Elicitation					
		First Response	Median	Second Response	Median		
Chamber availability: Not available due to low or high Pool. Consequences	Condition: The lock is not available when the tow arrives and requests lockage due to closure for repairs or maintenance. Please answer the following questions considering all tows that arrive to find maintenance or repairs:						
	Given that the chamber is closed for maintenance or repairs due to ordinary operation and maintenance, consider the impacts to industry:	8 2 6		8 2 6			
	1. What gr action of the total number of tows experience a short delay (<24 hrs)?*	2 2 2		2 2 2			
	1a How many hours would a typical tow be delayed?	4 3		4 3			
	2. What fraction of the total number of tows experience a long delay (>24 hrs)?	1 2 1.5	1 2 8	1 2 1.5		1 2 8	Low Median High
	2a How many days would a typical tow be delayed?	2		2			

*Only answers to question 1a are provided herein. Ayyub et al., 1996.

7.3 Economic consequences of floods

7.3.1 Background

Ayyub and Moser (2000) documented the use of methods and results of using expert-opinion elicitation for developing structural and content depth-damage relationships for single-family one-story homes without basements, residential content-to-structure value ratios (CSVR), and vehicle depth-damage relationships in the Feather River Basin of California. These damage functions consider exterior building material such as brick, brick veneer, wood frame, and metal siding. The resulting consequences can be used in risk studies and in making risk-based decisions.

The expert-opinion elicitation was performed during a face-to-face meeting of members of an expert panel developed specifically for the issues under consideration. The meeting of the expert panel was conducted after providing the experts, in advance of the meeting, background information, objectives, list of issues, and anticipated outcomes from the meeting. In Ayyub and Moser (2000), additional details on the different components of the expert elicitation process are described, the process itself is outlined and discussed, and the results are documented.

7.3.2 The Feather River Basin

7.3.2.1 Levee failure and consequent flooding

In January 1997, the eastern levee of the Feather River failed, causing major flooding near the Yuba County town of Arboga. Floodwaters inundated approximately 12,000 acres and damaged over 700 structures. Although the area was primarily agricultural, approximately 600 residential structures were affected by flooding. This area had a wide range of flooding depths, from maximum depths about 20 feet (structures totally covered) in the south near the levee break to minimal depths. Residential damage from the flooding was documented as a joint project of the Corps of Engineers Flood Damage Data Collection and the Sacramento-San Joaquin River Basin Comprehensive Study. The population of homes within the flood plain of the January 1997 flood defines the study area in this investigation.

7.3.2.2 Flood characteristics

The January 1997 flooding resulted from a trio of subtropical storms. Over a 3-day period, warm moist winds from the southwest, blowing over the Sierra Nevada, poured more than 30 inches of rain onto watersheds that were already saturated by one of the wettest Decembers on record. The first of the storms hit Northern California on December 29, 1996, with less than expected precipitation totals. Only 0.24 inch of rainfall was reported in Sacramento. On December 30, 1996, the second storm arrived. The third and most severe storm hit late December 31, 1996, and lasted through January 2, 1997.

Precipitation totals at lower elevations in the central valley were not unusually high, in contrast to extreme rainfall in the upper watersheds. Downtown Sacramento, for example, received 3.7 in of rain from December 26, 1996, through January 2, 1997. However, Blue Canyon (elevation 5000 ft) in the American River Basin received over 30 in of rainfall, thus providing for an orographic ratio of 8 to 1. A typical storm for this region would yield an orographic ratio of between 3 or 4 between these two locations.

In addition to the trio of subtropical storms, snowmelt also contributed to the already large runoff volumes. Several days before Christmas 1996, a cold storm from the Gulf of Alaska brought snow to low elevations in the Sierra Nevada foothills. Blue Canyon, for example, had a snowpack with 5 in of water content. The snowpack at Blue Canyon, as well as the snowpack at lower elevations, melted when the trio of warmer storms hit. Not much snowpack loss was observed, however, at snow sensors over 6000 ft in elevation in the northern Sierra. The effect of the snowmelt was estimated to contribute approximately 15 percent to runoff totals.

Prior to the late December storms, rainfall was already well above normal in the Sacramento River Basin. In the northern Sierra, total December precipitation exceeded 28 in, the second wettest December on record, exceeded only by the 30.8 in in December 1955.

On the Yuba River, available storage in New Bullards Reservoir was over 200 percent of flood management reservation space on December 1, 1996. By the end of the storm, available space was about 1 percent of flood pool. Oroville Reservoir, on the Feather River, began December with just over 100% flood management reservation space. At the completion of the storms in early January, approximately 27% space remained available.

The hydrologic conditions of the January, 1997 flooding of the Feather River Basin were used as the basis for developing depth-damage relationships and CSVR. These hydrologic conditions resulted in high-velocity flooding from an intense rainfall and a levee failure. A scenario that gives further details on flood characteristics was defined and used in the study.

7.3.2.3 Building characteristics

Most of the residential properties affected by the January 1997 flood were single-story, single-family structures with no basements. The primary construction materials were wood or stucco. Few properties in the study area were two-story, and nearly none had basements. It may be useful to differentiate one-story on slab from one-story on raised foundations.

The study is limited to the following residential structural types without basement:

One-story on slab,
One-story on piers and beams (i.e., raised foundations), and
Mobile homes.

7.3.2.4 Vehicle characteristics

Vehicle classes included in the study were

> Sedan cars,
> Pickup trucks, sport utility vehicles, vans, and
> Motorcycles.

7.3.3 Example issues and results

7.3.3.1 Structural depth-damage relationships

Background: The hydrologic conditions of the January 1997 flooding of the Feather River Basin were used as the basis for developing these values. These hydrologic conditions resulted in high-velocity flooding from an intense rainfall and a levee failure.

Issues: What are the best estimates of the median percent damage values as a function of flood depth to a residential structure for all types? Also, what is the confidence level in the opinion of the expert (low, medium, or high)?

The study is limited to the following residential structural types:

- Type 1 — one-story without basement on slab
- Type 2 — one-story on piers and beams (i.e., raised foundation)
- Type 3 — mobile homes

The experts discussed the issues that produced the assumptions provided in Table 7.7. In this study, structural depth-damage relationships were developed based on expert opinions as provided in sample results in Table 7.8. Each expert needed to provide his/her best estimate of the median value for percent damage and respondents' levels of confidence in their estimates. Sample revised depth damage relationships are shown in Figures 7.2a and 7.2b.

7.3.3.2 Content depth-damage relationships

Background: The hydrologic conditions of the January 1997 flooding of the Feather River Basin were used as the basis for developing these values. These hydrologic conditions resulted in high-velocity flooding from an intense rainfall and a levee failure.

Issues: What are the best estimates of the median percent damage values as a function of flood depth to the content of residential structures for all types? Also, what is the confidence level in the opinion of the expert (low, medium, or high)?

The study is limited to the following residential structural types:

- Types 1 and 2 — one-story without basement on slab or one-story on piers and beams (i.e., raised foundation)
- Type 3 — mobile homes

Chapter seven: Applications of expert-opinion elicitation 265

Table 7.7 Summary of Supportive Reasoning and Assumptions by Experts for Structure Value

House Types 1 and 2	House Type 3
Median house size of 1400 SF	Median size of 24 ft by 60 ft (1200 SF)
Wood frame homes	Wood frame homes
Median house value of $90,000 with land	Median house value of $30,000 without land
Median land value of $20,000	Median house age of 8 years
Median price without land is about $50 per SF	Finished floor is 3 ft above ground level
Median house age of 8 years	8 ft ceiling height
Type 2 has HVAC and sewer lines below finished floor	HVAC and sewer lines below finished floor
Percentages are of depreciated replacement value of houses	Percentages are of depreciated replacement value of houses
Flood without flow velocity	Flood without flow velocity
Several days of flood duration	Several days of flood duration
Flood water is not contaminated but has sediment without large debris	Flood water is not contaminated but has sediment without large debris
No septic field damages	No septic field damages
Allow for cleanup cost	Allow for cleanup cost

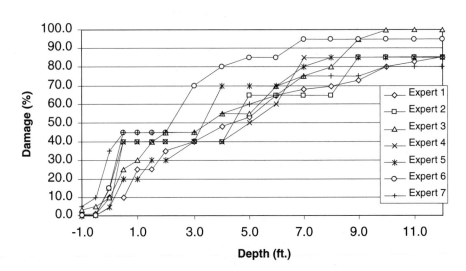

Figure 7.2a Percent damage to a residential structure Type 1: one-story without basement on slab.

The experts discussed the issues that produced the assumptions provided in Table 7.9. In this study, content depth-damage relationships were developed based on expert opinions as provided in the sample Table 7.10. Sample revised depth damage relationships are shown in Figures 7.3a and 7.3b.

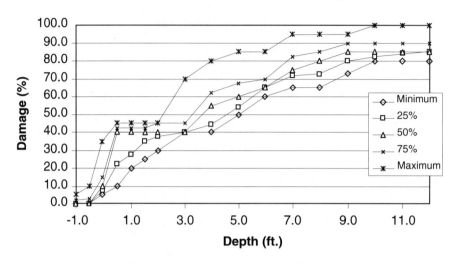

Figure 7.2b Aggregated percent damage to a residential structure Type 1: one-story without basement on slab.

7.3.3.3 Content-to-structure value ratios

Background: The hydrologic conditions of the January 1997 flooding of the Feather River Basin were used as the basis for developing these ratios. These hydrologic conditions resulted in high-velocity flooding from an intense rainfall and a levee failure.

Issues: What are the best estimates of the median values of a residential structure, its content, and their ratios (CSVR) for all types? Also, what is the confidence level in the opinion of the expert (low, medium, or high)?

The study is limited to the following residential structural types:

- Types 1 and 2 — one-story without basement on slab or one-story on piers and beams (i.e., raised foundation)
- Type 3 — mobile homes

The experts discussed the issues that produced the assumptions provided in Table 7.11. In this study, the best estimates of the median value of structures, the median value of contents, and the ratio of content to structure value were developed for these types based on expert opinions as provided in sample Table 7.12. Sample CVSRs are shown in Figure 7.4.

7.3.3.4 Vehicle depth-damage relationships

Background: The hydrologic conditions of the January 1997 flooding of the Feather River Basin were used as the basis for developing these values. These hydrologic conditions resulted in high-velocity flooding from an intense rainfall and a levee failure.

Chapter seven: Applications of expert-opinion elicitation   267

Table 7.8 Percent Damage to a Residential Structure Type 1: One-Story without Basement on Slab

Depth	Initial Estimate: % Damage by Expert							Min	Aggregated Opinions			Max
	1	2	3	4	5	6	7		25%	50%	75%	
−1	4	0	3	0	0	0	0	0	0	0	1.5	4
−0.5	4	0	5	0	0	0	0	0	0	0	2	5
0	5	0	10	5	0	10	0	0	0	5	7.5	10
0.5	10	40	12	7	10	13	45	7	10	12	26.5	45
1	15	40	25	9	20	15	55	9	15	20	32.5	55
1.5	20	40	28	11	30	20	55	11	20	28	35	55
2	30	40	35	13	30	20	60	13	25	30	37.5	60
3	40	40	35	15	40	30	60	15	32.5	40	40	60
4	48	40	40	25	70	50	65	25	40	48	57.5	70
5	53	65	40	40	70	85	70	40	46.5	65	70	85
6	65	65	45	50	70	85	75	45	57.5	65	72.5	85
7	68	70	75	70	80	90	75	68	70	75	77.5	90
8	70	75	80	90	80	90	75	70	75	80	85	90
9	73	85	95	100	95	90	75	73	80	90	95	100
10	80	85	100	100	100	100	80	80	82.5	100	100	100
11	83	85	100	100	100	100	80	80	84	100	100	100
12	85	85	100	100	100	100	80	80	85	100	100	100

Table 7.8 cont.

Depth	Revised Estimate: % Damage by Expert							Aggregated Opinions					
	1	2	3	4	5	6	7	Min	25%	50%	75%	Max
-1	1	0	3	0	0	0	5	0	0	0	2	5
-0.5	1	0	5	0	0	0	10	0	0	0	3	10
0	10	15	10	5	5	15	35	5	7.5	10	15	35
0.5	10	40	25	40	20	45	45	10	22.5	40	42.5	45
1	25	40	30	40	20	45	45	20	27.5	40	42.5	45
1.5	25	40	40	40	30	45	45	25	35	40	42.5	45
2	35	40	45	40	30	45	45	30	37.5	40	45	45
3	40	40	45	40	40	70	45	40	40	40	45	70
4	48	40	55	40	70	80	55	40	44	55	62.5	80
5	53	65	55	50	70	85	60	50	54	60	67.5	85
6	65	65	70	60	70	85	65	60	65	65	70	85
7	68	65	75	85	80	95	75	65	71.5	75	82.5	95
8	70	65	80	85	85	95	75	65	72.5	80	85	95
9	73	85	95	85	85	95	75	73	80	85	90	95
10	80	85	100	85	85	95	80	80	82.5	85	90	100
11	83	85	100	85	85	95	80	80	84	85	90	100
12	85	85	100	85	85	95	80	80	85	85	90	100
Confidence	High	High	High	High	High	High	High					

Chapter seven: Applications of expert-opinion elicitation 269

Table 7.9 Summary of Supportive Reasoning and Assumptions by Experts for Content Value

House Types 1, 2 and 3
As a guide, the insurance industry uses 70% ratio for the content to structure value
Median house value of $90,000 with land
Median land value of $20,000
Garage or shed contents are included
Median content age of 8 years
Percentages are of depreciated replacement value of contents
Flood without flow velocity
Several days of flood duration
Flood water is not contaminated but has sediment without large debris
Allow for cleanup cost
Insufficient time to remove (i.e. protect) contents

Issues: What are the best estimates of the median percent damage values as a function of flood depth to vehicles for all types? Also, what is the confidence level in the opinion of the expert (low, medium, or high)?

The study is limited to the following residential vehicle classes as follows:

- Type 1 — sedan cars
- Type 2 — pickup trucks, sports utility vehicles, and vans
- Type 3 — motorcycles

The experts discussed the issues that produced the assumptions provided in Table 7.13. In this study, the best estimates of the median value of vehicle depth-damage relationships were developed based on expert opinions as provided in sample Table 7.14. Sample relationships are shown in Figures 7.5a and 7.5b.

Table 7.10 Percent Damage to Contents of Residential Structure Types 1 and 2: One-Story on Slab or on Piers and Beams

Depth	Initial Estimate: % Damage by Expert							Aggregated Opinions				
	1	2	3	4	5	6	7	Min	25%	50%	75%	Max
-1	0.5	0	3	0	0	10	0	0	0	0	1.8	10
-0.5	0.5	0	5	0	0	20	0	0	0	0	2.8	20
0	2	30	15	0	0	40	5	0	1	5	22.5	40
0.5	2	40	35	20	50	40	10	2	15	35	40	50
1	15	50	35	40	50	40	20	15	27.5	40	45	50
1.5	27	60	40	50	60	40	20	20	33.5	40	55	60
2	35	70	40	60	70	60	40	35	40	60	65	70
3	47	80	70	70	80	80	40	40	58.5	70	80	80
4	55	80	70	80	80	90	60	55	65	80	80	90
5	80	80	70	90	90	90	60	60	75	90	90	90
6	90	80	70	100	100	90	85	70	82.5	90	95	100
7	90	80	75	100	100	95	95	75	85	95	97.5	100
8	90	85	85	100	100	100	100	85	87.5	100	100	100
9	90	85	90	100	100	100	100	85	90	100	100	100
10	90	85	90	100	100	100	100	85	90	100	100	100
11	90	85	90	100	100	100	100	85	90	100	100	100
12	90	90	90	100	100	100	100	90	90	100	100	100

Chapter seven: Applications of expert-opinion elicitation 271

Table 7.10 cont.

Depth	Revised Estimate: % Damage by Expert							Aggregated Opinions				
	1	2	3	4	5	6	7	Min	25%	50%	75%	Max
-1	2	0	3	0	0	2	0	0	0	0	2	3
-0.5	2	0	5	5	0	5	0	0	0	2	5	5
0	15	20	15	10	10	30	5	5	10	15	17.5	30
0.5	20	30	35	20	30	40	20	20	20	30	32.5	40
1	25	50	35	40	45	40	20	20	30	40	42.5	50
1.5	25	60	40	50	60	40	30	25	35	40	55	60
2	30	70	40	60	70	60	40	30	40	60	65	70
3	40	80	70	70	75	80	40	40	55	70	77.5	80
4	50	80	70	80	80	90	60	50	65	80	80	90
5	50	80	70	90	90	90	60	50	65	80	90	90
6	85	80	70	95	90	90	70	70	75	85	90	95
7	90	80	75	95	90	95	100	75	85	90	95	100
8	90	85	85	95	90	95	100	85	87.5	90	95	100
9	90	85	90	95	90	95	100	85	90	90	95	100
10	90	85	90	95	90	95	100	85	90	90	95	100
11	90	85	90	95	90	95	100	85	90	90	95	100
12	90	85	90	95	90	95	100	85	90	90	95	100
Confidence	High	High	High	High	High	High	High					

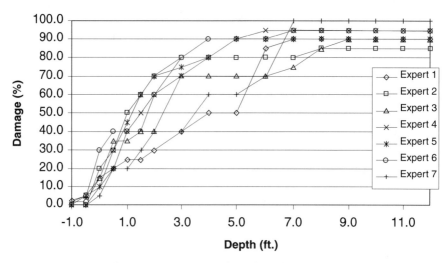

Figure 7.3a Percent damage to contents of residential structure Types 1 and 2: one-story on slab or on piers and beams.

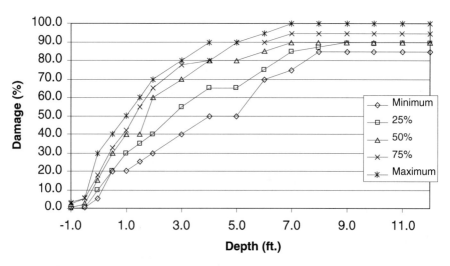

Figure 7.3b Aggregated percent damage to contents of residential structure Types 1 and 2: one-story on slab or on piers and beams.

Chapter seven: Applications of expert-opinion elicitation 273

Table 7.11 Summary of Supportive Reasoning and Assumptions by Experts for Content to Structure Value Ratio (CSVR)

House Types 1, 2 and 3
As a guide, the insurance industry uses 70% ratio for the content to structure value
Median house value of $90,000 with land
Median land value of $20,000
Garage or shed contents are included
Median content age of 8 years
Use depreciated replacement value of structure and contents
Insufficient time to remove (i.e. protect) contents

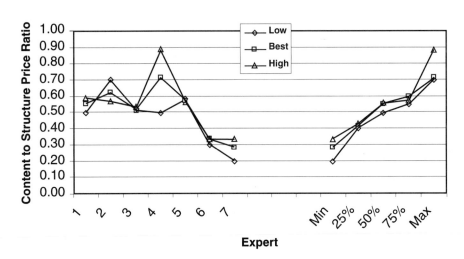

Figure 7.4 Content to structure value ratios (CSVR) for Types 1 and 2: one-story on slab or on piers and beams.

Table 7.12 Value of Residential Structures, Contents, and Their Ratios (CSVR) for Types 1 and 2: One-Story on Slab or on Piers and Beams

Issue	Initial Estimate: % Damage by Expert							Aggregated Opinions				
	1	2	3	4	5	6	7	Min	25%	50%	75%	Max
Median Structure (K$)												
Low	70	70	65	50	60	50	40	40	50	60	67.5	70
Best	90	110	106	70	70	60	70	60	70	70	98	110
High	110	250	175	90	80	80	90	80	85	90	142.5	250
Median Content (K$)												
Low	35	49	35	25	35	15	10	10	20	35	35	49
Best	50	77	41	50	40	20	20	20	30	41	50	77
High	65	175	70	80	45	25	25	25	35	65	75	175
CSVR												
Low	0.5	0.7	0.54	0.5	0.58	0.3	0.25	0.25	0.4	0.58	0.52	0.7
Best	0.56	0.7	0.39	0.71	0.57	0.33	0.29	0.33	0.43	0.59	0.51	0.7
High	0.59	0.7	0.4	0.89	0.56	0.31	0.28	0.31	0.41	0.72	0.53	0.7

Table 7.12 cont.

Issue	Revised Estimate: % Damage by Expert							Min	Aggregated Opinions			Max
	1	2	3	4	5	6	7		25%	50%	75%	
Median Structure (K$)												
Low	70	70	77	50	60	50	50	50	50	60	70	77
Best	90	80	82	70	70	60	70	60	70	70	81	90
High	110	90	94	90	80	75	90	75	85	90	92	110
Median Content (K$)												
Low	35	49	40	25	35	15	10	10	20	35	37.5	49
Best	50	50	42	50	40	20	20	20	30	42	50	50
High	65	51	50	80	45	25	30	25	37.5	50	58	80
CSVR												
Low	0.5	0.7	0.52	0.5	0.58	0.3	0.2	0.2	0.4	0.5	0.55	0.7
Best	0.56	0.63	0.51	0.71	0.57	0.33	0.29	0.29	0.42	0.56	0.6	0.71
High	0.59	0.57	0.53	0.89	0.56	0.33	0.33	0.33	0.43	0.56	0.58	0.89
Confidence	High	High	Medium	High	High	High	High					

Table 7.13 Summary of Supportive Reasoning and Assumptions by Experts for Vehicle Damage

Vehicle Types 1 and 2
Median vehicle age of 5 years
Percentages are of depreciated replacement value of vehicles
Flood without flow velocity
Several days of flood duration
Flood water is not contaminated but has sediment without large debris
Allow for cleanup cost

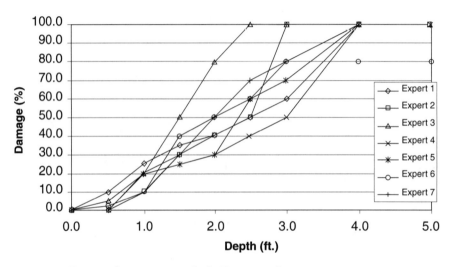

Figure 7.5a Percent damage to a vehicle Type 1: sedan cars.

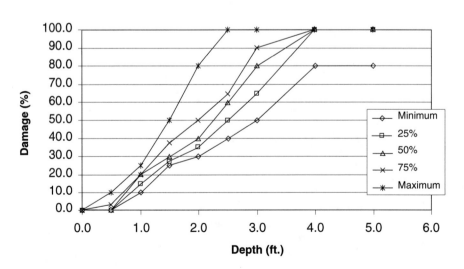

Figure 7.5b Aggregated percent damage to a vehicle Type 1: sedan cars.

Table 7.14 Percent Damage to a Vehicle Type 1: Sedan Cars

Depth	Initial Estimate: % Damage by Expert							Min	Aggregated Opinions			
	1	2	3	4	5	6	7		25%	50%	75%	Max
0	0	0	0	0	0	0	0	0	0	0	0	0
0.5	5	0	5	0	0	0	0	0	0	0	2.5	5
1	20	0	30	10	25	5	10	0	7.5	10	22.5	30
1.5	25	0	50	15	25	15	50	0	15	25	37.5	50
2	35	30	80	20	30	20	60	20	25	30	47.5	80
2.5	50	35	100	40	70	40	70	35	40	50	70	100
3	60	40	100	50	70	60	90	40	55	60	80	100
4	100	40	100	100	80	80	100	40	80	100	100	100
5	100	50	100	100	95	80	100	50	87.5	100	100	100

Depth	Revised Estimate: % Damage by Expert							Min	Aggregated Opinions			
	1	2	3	4	5	6	7		25%	50%	75%	Max
0	0	0	0	0	0	0	0	0	0	0	0	0
0.5	10	0	5	0	0	2	0	0	0	0	3.5	10
1	25	10	20	20	20	10	20	10	15	20	20	25
1.5	35	30	50	25	25	40	30	25	27.5	30	37.5	50
2	40	40	80	30	30	50	50	30	35	40	50	80
2.5	50	50	100	40	60	60	70	40	50	60	65	100
3	60	100	100	50	70	80	80	50	65	80	90	100
4	100	100	100	100	100	80	100	80	100	100	100	100
5	100	100	100	100	100	80	100	80	100	100	100	100
Confidence	High	High	High	High	High	Medium	High					

Bibliography

Abraham, D.M., Bernold, L.E., and Livingston, E.E., 1989. Emulation for control system analysis in automated construction. *J. Comput. Civ. Eng.*, 3(4), 320–332.

Alpert, M. and Raiffa, H., 1982, A progress report on the training of probability assessors, in Kahneman et al., (Eds.), *Judgement Under Uncertainty, Heuristics and Biases*, Cambridge University Press, Cambridge, 294–306.

Alvi, I.A. and Ayyub, B.M., 1990, Analysis of system performance under stochastic and fuzzy uncertainty, *NAFIPS*.

Amendola, A., 1986. *System Reliability Benchmark Exercise*, I and II, EUR-10696, EN/I, Joint Research Center of Ispra, Italy.

American Institute of Aeronautics and Astronautics, 1998. Guide for Verification and Validation of Computational Fluid Dynamics Simulation, AIAA G-077-1998.

American Psychological Association, 1985. Standards for Educational and Psychological Testing, Washington, D.C.

ASME (American Society of Mechanical Engineers), 1993. The use of decision-analytic reliability methods in codes and standards work. *CRTD* 23, New York.

Ammermn, R.R. Ed. 1970. *Belief, Knowledge and Truth*, Charles Scribners Sons, New York.

Ang, A. and Tang, W., 1975. *Probability Concepts in Engineering Planning and Design*, Vol. 1, John Wiley, New York.

Armstrong, D., 1973. *Belief Truth and Knowledge*, Cambridge University Press, Cambridge.

Austin, D.F. (Ed.), 1998. *Philosophical Analysis: A Defense by Example*, (Philosophical Studies Series, Vol. 39), D. Reidel Publishing Co.

Ayyub, B.M., 1994. The nature of uncertainty in structural engineering, in *Uncertainty Modelling and Analysis: Theory and Applications*, Ayyub and Gupta (Eds.), North-Holland-Elsevier Scientific Publishers, 195–210.

Ayyub, B.M., 1991. Systems framework for fuzzy sets in civil engineering, *Int. J. Fuzzy Sets and Syst*, 40(3), 491–508.

Ayyub, B.M., 1992. Generalized treatment of uncertainties in structural engineering. *Analysis and Management of Uncertainty: Theory and Applications*, Ayyub and Gupta, (Eds.) Elsevier Science Publisher, New York, 235–246.

Ayyub, B.M., 1992. Fault Tree Analysis of Cargo Elevators Onboard Ships, BMA Engineering Report, prepared for Naval Sea System Command, U.S. Navy, Crystal City, VA.

Ayyub, B.M., 1993. Handbook for Risk-Based Plant Integrity, BMA Engineering Report, prepared for Chevron Research and Technology Corporation, Richmond, CA.

Ayyub, B.M. and Chao, R.-J., 1998. Uncertainty modeling in civil engineering with structural and reliability applications, in *Uncertainty Modeling and Analysis in Civil Engineering*, B. Ayyub (Ed.), CRC Press, 1–32, Chap. 1.

Ayyub, B.M. and Gupta, M.M., (Eds.), 1997. *Uncertainty Analysis in Engineering and the Sciences: Fuzzy Logic, Statistics, and Neural Network Approach*, Kluwer Academic Publisher.

Ayyub, B.M. and Hassan, M.H.M., 1992a. Control of construction activities: I. Systems identification, *Civ. Eng. Sys.*, 9, 123–146.

Ayyub, B.M. and Hassan, M.H.M., 1992b. Control of construction activities: II. Condition assessment of attributes, *Civ. Eng. Sys.*, 9, 147–204.

Ayyub, B.M. and Hassan, M.H.M., 1992c. Control of construction activities: III. A fuzzy-based controller, *Civ. Eng. Sys.*, 9, 275–297.

Ayyub, B.M. and Lai, K.-L., 1992. Structural reliability assessment with ambiguity and vagueness in failure, *Nav. Eng. J.*, 104(3), 21–35.

Ayyub, B.M. and McCuen, R., 1996. *Numerical Methods for Engineers*, Prentice Hall, Upper Saddle River, NJ.

Ayyub, B.M. and McCuen, R., 1997. *Probability, Statistics and Reliability for Engineers*, CRC Press, Boca Raton, FL, 1997.

Ayyub, B.M. and Moser, D.A., 2000. Economic Consequence Assessment of Floods in the Feather River Basin of California Using Expert-Opinion Elicitation, Technical Report, Institute for Water Resources, USACE.

Ayyub, B.M., Gupta, M.M., and Kanal, L.N., 1992, *Analysis and Management of Uncertainty: Theory and Applications*, Elsevier Science Publishing Company, Inc., New York.

Ayyub, B.M., Riley, B.C., and Hoge, M.T., 1996. Expert Elicitation of Unsatisfactory-Performance Probabilities and Consequences for Civil Works Facilities, Technical Report, USACE, Pittsburgh District, PA.

Ayyub, B.M. and Haldar, A., 1984, Practical structural reliability techniques, *J. Struct. Eng.*, 110(8), 1707–1724.

Ayyub, B.M., (Ed.), 1998, *Uncertainty Modeling and Analysis in Civil Engineering*, CRC Press, Boca Raton, FL.

Ayyub, B.M., Guran, A., and Haldar, A., (Eds.), 1997. *Uncertainty Modeling in Vibration, Control, and Fuzzy Analysis of Structural Systems*, World Scientific.

Bailey, K.D., 1994. *Methods of Social Research*, The Free Press, Maxwell Macmillan, New York.

Beacher, G., Expert Elicitation in Geotechnical Risk Assessment, 1999. USACE Draft Report, University of Maryland, College Park, MD.

Beck, L.J., 1965. *The Metaphysics of Descartes*, Oxford University Press, Oxford, UK.

Bell, T.E. and Esch, K., 1989. The space shuttle: A case of subjective engineering, *IEEE Spectrum*, 42–46.

Bennett, J., 1990. Why is belief involuntary, *Analysis*, 50(2).

Blair, A.N. and Ayyub, B.M., 1999. Fuzzy stochastic cost and schedule risk analysis: MOB case study, *Proc. Symp. Very Large Floating Structures*, Elsevier, North Holland.

Blanchard, B.S., 1998. *System Engineering Management*, 2nd ed., John Wiley and Sons, Inc., New York.

Blockley, D.I., 1975, Predicting the likelihood of structural accidents, *Proc. Inst. Civ. Eng.*, 59, Part 2, 659–668.
Blockley, D.I., 1979a, The calculations of uncertainty in civil engineering, *Proc. Inst. Civ. Eng.*, 67, Part 2, 313–326.
Blockley, D.I., 1979b, The role of fuzzy sets in civil engineering, *Fuzzy Sets and Syst.*, 2, 267–278.
Blockley, D.I., 1980, *The Nature of Structural Design and Safety*, Ellis Horwood, Chichester, UK.
Blockley, D.I., Pilsworth, B.W., and Baldwin, J.F., 1983, Measures of Uncertainty, *Civ. Eng. Syst.*, 1, 3–9.
Boltzmann, L., 1894, Zur Integration der Diffusionsgleichung bei variablen Diffusionskoeffizienten, *Ann. Physik*, 53, Leipzig, Germany.
Bouissac, P., 1992. The construction of ignorance and the evolution of knowledge, *University of Toronto Quarterly*, 61(4), Summer.
Bowles, D., 1990. Risk assessment in dam safety decisionmaking. *Risk-Based Decision Making in Water Resources*, Proc. 4th Conf., Y. Y. Haimes and E.Z. Stakhiv (Eds.), 254–83.
Bradie, M., 1986. Assessing evolutionary epistemology, *Biol. and Philos.*, 1, 401–450.
Bremermann, H.J., 1962. Optimization through evolution and recombination, in Yovits, M.C., Jacobi, G.T., and Goldstein, G.D. (Eds.), *Self-Organizing Systems, Spartan Books*, Washington, 93–106.
Brown, C.B. and Yao, J.T.P., 1983, Fuzzy sets and structural engineering, *J. Struct. Eng.*, 109(5), 1211–1225.
Brown, C.B., 1979, A fuzzy safety measure, *J. Eng. Mech. Div.*, 105(EM5), 855–872.
Brown, C.B., 1980, The merging of fuzzy and crisp information, *J. Eng. Mech. Div.*, 106(EM1), 123–133.
Brune, R., Weinstein, M., and Fitzwater, M., 1983. Peer Review Study of the Draft Handbook for Human Reliability Analysis with Emphasis on Nuclear Power Plant Applications, NUREG/CR-1278.
Campbell, D.T., 1974. Evolutionary epistemology, in P.A. Schilpp (Ed.), *The Philosophy of Karl Popper*, Open Court Publishing, LaSalle, IL, 1,413–463.
Chestnut, H., 1965. *Systems Engineering Tools*, John Wiley & Sons, Inc., New York.
Chomsky, N., 1968. *Language and Mind*, Harcourt Brace Jovanovich, New York.
Clemen, R.T., 1989. Combining forecasts: a review and annotated bibliography, *Int. J. Forecasting*, 5, 559–583.
Colglazier, E.W. and Weatherwax, R.K., 1986. Failure estimates for the space shuttle, Abstracts from the Society of Risk Analysis, annual meeting, Boston, MA, Nov. 9–12, 1986, 80.
Committee on Safety Criteria for Dams, 1985. Safety of Dams: Flood and Earthquake Criteria, National Academy Press, Washington, D.C.
Committee on the Safety of Existing Dams, 1983. Safety of Existing Dams, Evaluation and Improvement, National Research Council, National Academy Press, Washington, D.C.
Cooke, R.M., 1986. Problems with empirical bayes, *Risk Analysis*, 6(3), 269–272.
Cooke, R.M., 1991. *Experts in Uncertainty*, Oxford University Press, New York.
Dancy, J., 1985. *Contemporary Epistemology*, Basil Blackwell, Oxford, UK.
De Cooman, G., 1997, Possibility theory. *Int. J. G. Sys.*, 25(4), 291–371.
De Finetti, B., 1937, English Translation in 1964 by H. Kyburg and H. Smokler (Eds.). *Studies in Subjective Probabilities*, Wiley, New York.

Defense Acquisition University, 1998. Risk Management Guide, Defense Systems Management College Press, Fort Belvoir, VA.

DeKay, M.L. and McClelland, G.H., 1993, Predicting loss of life in cases of dam failure and flash flood, *Risk Analysis*, 13(2), 193–205.

Dempster, A.P., 1976a. Upper and lower probabilities induced by multivalued mapping, *Ann. Math. Stat.*, 38(2), 325–339.

Dempster, A.P., 1976b. Upper and lower probabilities inferences based on a sample from a finite univariate population, *Biometrika*, 54(3–4), 515–528.

Dewey, J., 1929. *The Quest For Certainty*, Minton, Balch and Co., New York.

di Carlo, C.W., 1998. Evolutionary Epistemology and the Concept of Ignorance, PhD Thesis, University of Waterloo, Ontario, Canada.

Dombi, J., 1982. A general class of fuzzy operators, the De Morgan class of fuzzy operators and fuzziness measures induced in fuzzy operators, *Fuzzy Sets and Systems*, 8, 149–163.

Dong, W-M. and Wong, F.S., 1986a, From uncertainty to approximate reasoning: Part 1: Conceptual models and engineering interpretations, *Civ. Eng. Sys.*, 3(3), 143–154.

Dong, W-M. and Wong, F.S., 1986b, From uncertainty to approximate reasoning: Part 2: Reasoning with algorithmic rules, *Civ. Eng. Sys.*, 3(4), 192–202.

Dong, W-M. and Wong, F.S., 1986c, From uncertainty to approximate reasoning: Part 3: Reasoning with conditional rules, *Civ. Eng. Sys.*, 3(5), 112–121.

Dretske, F., 1981. *Knowledge and the Flow of Information*, MIT Press, Cambridge, MA.

Driver, J., 1989. The virtues of ignorance, *J. Philos.*, 86.

Dubois, D. and Prade, H., 1988. *Possibility Theory*. Plenum Press, New York.

Dubois, D. and Prade, H., 1980. On several representation of an uncertain body of evidence, in *Fuzzy Information and Decision Processes*, Gupta, and Sanchez (Eds.), North-Holland, New York, 167–181.

Duncan, R. and Weston-Smith, M., Eds. 1977. *The Encyclopedia of Ignorance*, Pergamon Press Ltd., New York.

Durant, W., 1991. *Story of Philosophy: The Lives and Opinions of the World's Greatest Philosophers*, Pocket Books.

Eldukair, Z.A. and Ayyub, B.M., 1991. Analysis of recent U.S. structural and construction failures, *J. Performance of Constructed Facilities*, 5(1), 57–73.

Ferrell, W.R., 1985. Combining individuals judgments, in Wright, G. (Ed.), *Behavioral Decision Making*, Plenum, NY.

Ferrell, W.R., 1994. Discrete subjective probabilities and decision analysis: elicitation, calibration and combination, in Wright, G., and Ayton, P. (Eds.), *Subjective Probability*, John Wiley and Sons, New York.

Ferson, S., Root, W., and Kuhn, R., 1999. RAMAS Risk Calc: Risk Assessment with Uncertain Numbers, Applied Biomathematics, Inc., Setauket, NY.

Fisher, R.A., 1930. *The Genetical Theory of Natural Selection*, Oxford University Press, Oxford, UK.

Frank, M.J., 1979. On the simultaneous associativity of $F(x,y)$ and $x+y-F(x,y)$, *Aequationes Mathematicae*, 19(2–3), 194–226.

Frank, M.J., Nelson, R.B., and Schweizer, B., 1987, Best-possible bounds for the distribution of a sum — a problem of Kolmogorov, *Probability Theory and Related Fields*, 74, 199–211.

Freeman, W.M., 1969. *Readings from Scientific America: Science, Conflict and Society*, San Francisco.

French, S., 1985. Group consensus probability distributions: a critical survey, J.M. Bernardo et al. (Eds.), *Bayesian Statistics*, Elsevier, North Holland, 183–201.

Furuta, H., Fu, K.S., and Yao, J.T.P., 1985, Structural engineering applications of expert systems, *Computer Aided Design*, 17(9), 410–419.

Furuta, H., Shiraishi, N., and Yao, J.T.P., 1986, An expert system for evaluation of structural durability, *Proc. 5th OMAE Symp.*, 1, 11–15.

Galanter, E., 1962. The direct measurement of utility and subjective probability, *Am. J. Psych.*, 75, 208–220.

Garey, M.R. and Johnson, D.S., 1979. *Computers and Intractability: A Guide to the Theory of NP-Completeness*, W.H. Freeman, San Francisco.

Genest, C. and Zidek, J., 1986. Combining probability distributions: critique and an annotated bibliography, *Stat. Sci.*, 1(1), 114–148.

Goldman, A., 1993. *Philosophical Applications of Cognitive Science*, Westview Press, Boulder, CO.

Gustafson, D.H., Shukla, R.K., Delbecq, A., and Walster, G.W., 1973. A comparative study of differences in subjective likelihood estimates made by individuals, interacting groups, delphi groups, and nominal groups, *Organizational Behavior and Human Performance* 9, 200–291.

Hackett, C., 1969. The origin of speech, *Scientific American* 203, 88–111.

Haldar, A., Guran, A., and Ayyub, B.M., (Eds.), 1997. *Uncertainty Modeling in Finite Element, Fatigue, and Stability of Systems*, World Scientific.

Hall, A.D., 1962. *A Method for Systems Engineering*. D. Van Nostrand Company, Inc., Princeton, NJ.

Hall, A.D., 1989. *Metasystems Methodology, A New Synthesis and Unification*. Pergamon Press, New York, NY.

Hall, J.W., Blockley, D.I., and Davis, J.P., 1998, Uncertain inference using interval probability theory, *Int. J. Approximate Reasoning*, 19(3,4), 247–264.

Hallden, S., 1986. *The Strategy of Ignorance: From Decision Logic to Evolutionary Epistemology*, Library of Theoria, Stockholm.

Harr, M.E., 1987. *Reliability-Based Design in Civil Engineering*, McGraw-Hill Book Company, New York.

Hartford, D.N.D., 1995, *How Safe is Your Dam? Is it Safe Enough?*, B.C. Hydro, Maintenance, Engineering, and Projects, Burnaby, BC.

Hartley, R.V., L., 1928, Transmission of information, *The Bell Systems Technical J.*, 7(3), 535–563.

Hassan, M.H.M. and Ayyub, B.M., 1994. Multi-attribute fuzzy control of construction activities, in *Uncertainty Modelling and Analysis: Theory and Applications*, Ayyub and Gupta, (Eds.), North-Holland-Elsevier Scientific Publishers, 271–286.

Hassan, M.H.M., Ayyub, B.M., and Bernold, L., 1992. Fuzzy–based real-time control of construction activities, *Analysis and Management of Uncertainty: Theory and Applications*, Ayyub, Gupta, and Kanal (Eds.), North-Holland-Elsevier Scientific Publishers, 331–350.

Hassan, M.H.M. and Ayyub, B.M., 1993a. Multi-attribute control of construction activities, *Civ. Eng. Sys.*, 10, 37–53.

Hassan, M.H.M. and Ayyub, B.M., 1993b. A fuzzy controller for construction activities, *Fuzzy Sets and Systems*, North-Holland, 56(3), 253–271.

Hassan, M.H.M. and Ayyub, B.M., 1997. Structural fuzzy control, in *Uncertainty Modeling in Vibration, Control and Fuzzy Analysis of Structural Systems*, Ayyub, Guran, and Haldar (Eds.), World Scientific, 179–232, Chap. 7.

Helmer, O., 1968. Analysis of the future: the delphi method, and the delphi method — an illustration, in J. Bright (Ed.), *Technological Forecasting for Industry and Government*, Prentice Hall, Englewood Cliffs, NJ.

Henley, E.J. and Kumamoto, H., 1981. *Reliability Engineering and Risk Assessment*, Prentice-Hall, Inc., Englewood Cliffs, NJ., 1981.

Higashi, M. and Klir, G.J., 1983, Measures of uncertainty and information based on possibility distributions, *Int. J. Gen. Sys.*, 8(3), 43–58.

Honderich, H. (Ed.), 1995. *The Oxford Companion to Philosophy*, Oxford University Press, New York.

Horgan, John, 1996. *The End of Science*. Addison-Wesley Publishing Co., Reading, MA.

Horgan, Terry (Ed.), 1995. Vagueness. The Spindel Conference 1994. *Southern J. Philos.*, 33 (Supplement).

Horwich, P., 1987. *Asymmetries in Time: Problems in the Philosophy of Science*, MIT Press, Cambridge, MA.

Ishizuka, M., Fu, K.S., and Yao, J.T.P., 1981, A rule-inference method for damage assessment, *Preprint 81–502*, St. Louis, MO.

Ishizuka, M., Fu, K.S., and Yao, J.T.P., 1983, Rule-based damage assessment system for existing structures, *Solid Mech. Arch.*, 8, 99–118.

Itoh, S. and Itagaki, H., 1989, Application of fuzzy-Bayesian analysis to structural reliability, *Proceedings 5th Int. Conf. Struct. Safety and Reliability*, 3, A. H-S. Ang, M. Shinozuka, and G.I. Schuëller (Eds.), 1771–1774.

Johnson, O.A., 1979. Ignorance and irrationality: a study in contemporary scepticism, *Philos. Res. Arch.*, 5,(13) 10.

Johnson-Laird, P., 1988. *The Computer and the Mind: An Introduction to Cognitive Science*. Harvard University Press, Cambridge, MA.

Kahn, H., 1960. *On Thermonuclear War*, Free Press, New York.

Kahn, H. and Wiener, A.J., 1967. *The Year 2000: A Framework for Speculation*, Macmillan, New York.

Kahneman, D., Slovic, P., and Tversky, A. (Eds.), 1982. *Judgment Under Uncertainty: Heuristics and Biases*. Cambridge University Press, Cambridge.

Kaneyoshi, M., Tanaka, H., Kamei, M., and Furuta, H., 1990, Optimum cable tension adjustment using fuzzy regression analysis, *3rd WG 7.5 Working Conference on Reliability and Optimization of Structural Systems, International Federation for Information Processing*, University of California, Berkeley, CA.

Kaufman, A. and Gupta, M.M., 1985. *Introduction to Fuzzy Arithmetic, Theory and Applications*, Van Nostrand Reinhold Co., New York, 1985.

Kaufmann, A., 1975. *Introduction to the Theory of Fuzzy Subsets*, Academic Press, New York, (Translated by D. L. Swanson).

Klir, G.J., 1969. *An Approach to General Systems Theory*. Van Nostrand Reinhold Company, New York.

Klir, G.J., 1999, On fuzzy-set interpretation of possibility theory, *Fuzzy Sets and Sys.*, 108(3), 263–273.

Klir, G.J., 1985, *Architecture of Systems Problem Solving*, Plenum Press, New York.

Klir, G.J. and Cooper, J.A., 1996. On constraint fuzzy arithmetic, *Proceedings 5th IEEE Int. Conf. Sys.*, 1693–1699.

Klir, G.J. and Folger, T.A., 1988. *Fuzzy Sets, Uncertainty, and Information*, Prentice Hall, New Jersey.

Klir, G.J. and Wierman, M.J., 1999. Uncertainty-based information: elements of generalized information theory. *Studies in Fuzziness and Soft Computing*, Physica-Verlag, New York.

Bibliography

Klir, G.J. and Yuan, B., 1995. *Fuzzy Sets, Fuzzy Logic, and Fuzzy Sets: Selected Papers by Lotfi Zadeh.* World Scientific, Singapore, Indonesia.

Krueger, R.A. and Casey, M.A., 2000. *Focus Groups: A Practical Guide for Applied Research.* 3rd ed., Sage Publication, Thousand Oaks, CA.

Kumamoto, H. and Henley, E.J., 1996. *Probabilistic Risk Assessment and Management for Engineers and Scientists,* 2nd ed., IEEE Press, New York.

Lai, K.-L. Lai, 1992. Generalized Uncertainty in Structural Reliability Assessment, PhD Dissertation, University of Maryland, College Park, MD.

Lai, K.-L. and Ayyub, B.M., 1994. Generalized uncertainty in structural reliability assessment, *Civ. Eng. Syst.,* 11(2), 81–110.

Langer, E., 1975. The illusion of control, *J. of Personality Soc. Psych.,* 32, 311–328.

Levi, I., 1977. Four types of ignorance, *Soc. Res.,* 44, Winter.

Lichtenstein, S. and Newman, J.R., 1967. Empirical scaling of common verbal phrases associated with numerical probabilities, *Psychonomic Science,* 9(10), 563–564.

Lindley, D., 1970. *Introduction to Probability and Statistics from a Bayesian Viewpoint,* Cambridge University Press, UK.

Linstone, H.A. and Turoff, M., 1975. *The Delphi Method, Techniques and Applications,* Addison Wesley, MA.

Merton, R.K., Fiske, M., and Kendall, P.L., 1956. *The Focused Interview,* Free Press, Glencoe, IL.

Miller, J.G., 1978. *Living Systems,* McGraw Hill 1978; University of Colorado, 1995.

Montagu, A. and Darling, E., 1970. *The Ignorance of Certainty.* Harper and Row Publishers, New York.

Moore, R.E., 1966. *Interval analysis.* Prentice-Hall, Inc., Englewood Cliffs, NJ.

Moore, R.E., 1979. *Methods and Applications of Interval Analysis.* SIAM, Philadelphia.

Morgan, M.G. and Henrion, M., 1992. *Uncertainty: A Guide to Dealing with Uncertainty in Quantitative Risk and Policy Analysis,* Cambridge University Press, New York.

Morris, J.M. and D'Amore, R.J., 1980. Aggregating and communicating uncertainty, Pattern Analysis and Recognition Corp., 228 Liberty Plaza, Rome, NY.

Murphy, A. and Daan, H., 1984. Impact of feedback and experience on the quality of subjective probability forecasts: comparison of the results from the first and second years of the Zierikzee Experiment, *Monthly Weather Review,* 112, 413–423.

Nelson, R.B., 1999. An introduction to couplas, *Lecture Notes in Statistics,* Vol. 139, Springer-Verlag, New York.

Newman, J.R., 1961. Thermonuclear war, *Scientific America,* March 1961.

Nuclear Regulatory Commission, 1975. Reactor Safety Study, WASH-1400, NUREG 751014.

Nuclear Regulatory Commission, 1997. Recommendations for Probabilistic Seismic Hazard Analysis: Guidance on Uncertainty and Expert Use, prepared by the Senior Seismic Hazard Analysis Committee, NUREG/CR-6372, UCRL-ID-122160, Vol. 1 and 2, Washington, D.C.

Pal, S.K. and Skowron, A. (Eds.) 1999. *Rough Fuzzy Hybridization.* Springer-Verlag, Singapore and New York.

Paté-Cornell, E., 1996. Uncertainties in risk analysis and comparability of risk estimates, Society for Risk Analysis 1996 Annual Meeting, McLean, VA.

Pawlak, Z., 1991. *Rough Sets: Theoretical Aspects of Reasoning About Data.* Kluwer, Boston.

Pawlak, Z., 1999. Rough sets, rough functions and rough calculus, in Pal and Skowron (eds.) *Rough Fuzzy Hybridization*, Springer-Verlag, Singapore, and New York.
Piattelli-Palmarini, M., 1994. *Inevitable Illusions: How Mistakes of Reason Rule our Minds*. John Wiley and Sons, New York (translated from Italian).
Ponce, V.M., 1989. *Engineering Hydrology Principles and Practices*, Prentice-Hall, Englewood Cliffs, NJ.
Popkin, R.H. (Ed.), 2000. *The Columbia History of Western Philosophy*. Mjf Books.
Preyssl, C. and Cooke, R., 1989. Expert judgment: subjective and objective data for risk analysis for space-flight systems, *Proc. PSA 1989*, Pittsburgh.
Ramsey, F., 1931. Truth and probability, in Braithwaite (Ed.), *The Foundation of Mathematics*, Kegan Paul, London, 156–198.
Regan, H.M., Ferson S., and Berleant, D., 2000, Equivalence of five methods for reliable uncertainty propagation, Draft paper, Applied Biomathematics, Inc., Setauket, NY.
Reichenbach, H., 1951. *The Rise of Scientific Philosophy*, University of California Press, California, 1968 edition.
Rice, S.A., (Ed.), 1931. *Methods in Social Research*, University of Chicago Press, Chicago, 561.
Rowe, G., 1992. Perspectives on expertise in aggregation of judgments, in Wright, G., and Bolger, F. (Eds.), *Expertise and Decision Support*, Plenum Press, New York, 155–180.
Russell, B., 1975. *A History of Western Philosophy*, Simon and Schuster, New York.
Sackman, H., 1975. *Delphi Critique: Expert Opinion, Forecasting and Group Process*, Lexington Books, Lexington, MA.
Samet, M.G., 1975. Quantitative interpretation of two qualitative scales used to rate military intelligence, *Human Factors*, 17(2), 192–202.
Schweizer, B. and Sklar, A., 1983. *Probability Metric Spaces*, North-Holland, New York.
Science, 1983. 222(4630), 1293.
Shafer, G., 1976. *A Mathematical Theory of Evidence*, Princeton University Press, Princeton, NJ.
Shannon, C.E., 1948. The mathematical theory of communication, *The Bell System Technical J.*, 27, 379–423, 623–656.
Shiraishi, N. and Furuta, H., 1983, Reliability analysis based on fuzzy probability, *J. Eng. Mech.*, 109(6), 1445–1459.
Shiraishi, N., Furuta, H., and Sugimoto, M., 1985, Integrity assessment of structures based on extended multi-criteria analysis, *Proc. 4th ICOSSAR*.
Smithson, M., 1985. Towards a social theory of ignorance, *J. Theory of Soc. Beh.*, 15, 151–172.
Smithson, M., 1988. *Ignorance and Uncertainty*, Springer-Verlag, New York.
Smithson, M., 1989. *Ignorance and Uncertainty*, Springer-Verlag, New York.
Sober, E. 1991. *Core Questions in Philosophy*, Macmillan Publishing Company, New York.
Solomon, R. and Higgins, K., 1996. *A Short History of Philosophy*, Oxford University Press, New York.
Spetzler, C.S. and Stael von Holstein, C-A.S., 1975. Probability encoding in decision analysis, *Management Science*, 22(3).
Stillings, N. et al., 1995. *Cognitive Science*. 2nd ed. MIT Press, Cambridge, MA.
Sugeno, M., 1974. Theory of Fuzzy Intervals and Its Applications. PhD Dissertation, Tokyo Institute of Technology, Tokyo.

Sugeno, M., 1977. Fuzzy measures and fuzzy integrals: a survey, in Gupta, M.M., Saridis, G.N. and Gaines, B.R. (Eds.), *Fuzzy Automata and Decision Processes*, North-Holland, Amsterdam and New York, 89–102.
Thagard, P., 1996. *Mind: Introduction to Cognitive Science*, MIT Press, Cambridge, MA.
Thys, W., 1987. Fault Management. Ph.D. Dissertation, Delft University of Technology, Delft.
Trauth, K.M., Hora, S.C. and Guzowski, R.V., 1993. Expert Judgement on Markers to Deter Inadvertent Human Intrusion into the Waste Isolation Pilot Plant, Report SAND92-1382, Sandia National Laboratories, Albuquerque, NM.
U.S. Army Corps of Engineers, 1965. Standard Project Flood Determinations. Civil Engineer Bulletin No. 52–8, Engineering Manual EM 1110–2–1411.
U.S. Army Corps of Engineers, 1982. National Program of Inspection of Nonfederal Dams, Final Report to Congress, Engineering Report ER 1110–2-106.
U.S. Army Corps of Engineers, 1997. Guidelines for Risk-based Assessment of Dam Safety, Draft Engineering Pamphlet EP 1110-1-XX, CECW-ED.
U.S. Bureau of Reclamation, *Policy and Procedures for Dam Safety Modifications*, USBR, Denver, 1989.
U.S. Interagency Advisory Committee on Water Data, Hydrology Subcommittee, 1982. Guidelines for Determining Flood Flow Frequency. Bulletin No. 17B, USGS, Reston, VA.
Unger, P., 1975. *Ignorance*. Clarendon Press, Oxford, UK.
von Eckardt, B., 1993. *What is Cognitive Science?* MIT Press, Cambridge, MA.
Waldrop, M.M., 1992. *Complexity: The Emerging Science at the Edge of Order and Chaos*, Simon and Schuster, New York.
Walley, P., 1991. *Statistical Reasoning with Imprecise Probabilities*, Chapman and Hall, London, UK.
Wang, Z. and Klir, G.J., 1992. *Fuzzy Measure Theory*, Plenum Press, New York.
Weaver, W., 1948. Science and complexity, *American Scientist*, 36(4), 536–544.
Webster's New World College Dictionary, 1995. Zane Publishing, Inc., and 1988 Simon and Schuster, Inc.
Weinstein, D. and Weinstein, M.A., 1978. The sociology of non-knowledge: a paradigm. in R.A. Jones (Ed.) *Research in the Sociology of Knowledge, Science and Art*, 1, JAI Press, New York.
White, G.J. and Ayyub, B.M., 1985, Reliability methods for ship structures, *Nav. Eng. J.*, 97(4), 86–96.
Wiggins, J., 1985. ESA Safety Optimization Study, Hernandez Engineering, HEI-685/1026, Houston.
Williams, B. 1995. Philosophy and the understanding of ignorance, *Diogenes*, 4311(169), Spring.
Williamson, R. and Downs, T., 1990, Probabilistic arithmetic I: numerical methods for calculating convolutions and dependency bounds, *Int. J. Approximate Reasoning*, 4, 89–158.
Williamson, T. 1994. *Vagueness*. Routledge.
Wilson, B., 1984. *Systems: Concepts, Methodologies, and Applications*. John Wiley and Sons, Inc., New York.
Winkler, R.N. and Murphy, A., 1968, Good probability assessors, *J. Applied Meteorology*, 7, 751–758.
Wuketits, F.M., 1990. *Evolutionary Epistemology and Its Implications for Humankind*, State University of New York Press, Albany, NY.

Yager, R.R., 1979. On the measure of fuzziness and negation. I: membership in the unit interval, *Int. J. Gen. Syst.*, 5(4), 189–200.

Yager, R.R., 1980a. On the general class of fuzzy connectives, *Fuzzy Sets and Syst.*, 4(3), 235–242.

Yager, R.R., 1980b. On the measure of fuzziness and negation. II: *Lattices, Information and Control*, 44(3), 236–260.

Yager, R.R., 1983, Entropy and specificity in a mathematical theory of evidence, *Int. J. Gen. Syst.*, 9, 249–260.

Yager, R.R., 1986, Arithmetic and other operations on Dempster-Shafer structures, *Int. J. Man-Machine Studies*, 25, 357–366.

Yao, J.T.P. and Furuta, H., 1986, Probabilistic treatment of fuzzy events in civil engineering, *Probabilistic Eng. Mech.*, 1(1), 58–64.

Yao, J.T.P., 1979, Damage assessment and reliability evaluation of existing structures, *Eng. Struct.*, England, 1, 245–251.

Yao, J.T.P., 1980, Damage assessment of existing structures, *J. Eng. Mech. Div.*, 106(EM4), 785–799.

Zadeh, L.A., 1965, Fuzzy sets, *Information and Control*, 8, 338–353.

Zadeh, L.A., 1968, Probability measures of fuzzy events, *J. Math. Analysis*, 23, 421–427.

Zadeh, L.A., 1973, Outline of a new approach to the analysis of complex systems and decision processes, *IEEE Trans. Syst. Man and Cybernetics*, SMC-3(1), 28–44.

Zadeh, L.A., 1975, The concept of linguistic variable and its application to approximate reasoning, Parts I, II and III, *Information and Control*, 8, 199–249, 301–357, 9, 43–80.

Zadeh, L.A., 1987, Fuzzy sets as a basis for a theory of possibility, *Fuzzy Sets and Syst.*, 1, 3–28.

Zadeh, L.A., Fu, K.S., Tanaka, K. and Shimara, J., 1975, *Fuzzy Sets and Their Application to Cognitive and Decision Processes*, Academic Press, New York.

Zimmerman, H.J., 1985. *Fuzzy Set Theory-and its Applications*. Kluwer-Nijhoff Publishing, Boston.

Index

A

absence, 31-33
absolute idealism, 15
absolute Spirit, 15
absolute truth, 17
abstraction, 68, 85, 88, 94
accountability, 110
accuracy rating, 109
addition, 130, 140
additive monotone measure, 163, 165
additive probability, 170
aesthetics, 4
aggregation, 147, 221
    opinion, 246
aggregation of expert opinions, 222
aggregation of opinions, 222, 231
AIAA, 116
AIDS, 91
Air Force, 100,
Al-Farabi, 5, 11, 13
Al-Ghazali, 12
Al-Kindi, 5, 11, 13
alpha cut, 134, 155
    strong, 135
alternative futures, 105
ambiguity, 31-33, 86
American Psychological Association (APA), 114
analysis, 116
analytical priori, 15
Anaximander, 3
Anaximenes, 3
anchoring, 106
Andronicus, 3
anthropology, 24
Antisthenes, 6
antithesis, 15

APA, see American Psychological Association
applications
    Arboga, 262-272
    Army Corps of Engineers, 262-272
    basin, 262-272
    brick, 262-272
    California, 262-272
    cargo, 250
    chamber, 251
    consequence, 256, 262-272
    content, 262-272
    content-to-structure value, 262-272
    CSVR, 262-272
    dam, 251
    damage, 262-272
    damage ratio, 262-272
    damage-depth ratio, 262-272
    elevators, 250
    Emsworth, 251
    Feather River, 262-272
    flood, 262-272
    floodwater, 262-272
    frame, 262-272
    houses, 262-272
    lock, 251
    miter gate, 256
    mooring, 256
    naval, 250
    navigation, 251
    Ohio River, 251
    rainfall, 262-272
    ships, 250
    siding, 262-272
    snow, 262-272
    storm, 262-272
    structure, 262-272
    tow, 256
    USACE, see Army Corps of Engineers

vehicle, 262-272
vehicle damage, 262-272
wood, 262-272
Yuba County, 262-272
approximate interval, 136
approximate number, 136
approximations, 31-32, 34
Aquinas, 11, 13
Arabic, 10, 12
Arboga 262-272
Aristotelian logic, 12
Aristotle, 3, 7, 9-10
arithmetic, 130
    constraint, 141
    nonconstraint, 141
    nonrestricted, 140
    restricted, 140
    vertex method, 140
Army Corps of Engineers, 56, 262-272
artificial intelligence, 22
assignment, 211, 213
associative laws, 131
atomic facts, 17
attributes, 68
Austin, 16, 19
availability, 106
average, 173
average time to failure, 176
averaging, 147
Averroes, 11-13
Avicenna, 11, 13
axiom probability, 172

B

Bacon, 12-13
basic assignment
    focal element, 167
    probability, 171
    rule of combination, 167
    strongest, 167
    weakest, 167
basin, 57, 262-272
Bayes theorem, 107, 182
Bayesian, 85-86
    methods, 181
    objective, 181
    posterior, 182
    prior, 182
    probabilities, 182
    reverse probability, 182
    subjective, 182
    total probability, 182
behaviorists, 24

behavior function, 76, 79
behavior system, 76
being theory, 7
belief, 98, 199, 213
    evidence theory, 165
    monotone measure, 164
    probability, 170
belief measure of evidence theory, 165
belonging, 128, 132
Berkeley, 14-15
beta, 220
betting, 112
betting rates, 112
bias, 116
binary, 210
bird species, 204, 231
bits, 92, 210
black-box method, 54
blind ignorance, 28, 30-31
boundaries, 135
bounds, 136
Bradley, 16, 19
Bremermann, 92
Bremermann limit, 92
brick, 262-272
bridge, 168
bridge failure, 168
byte, 210

C

calibration, 107, 111
California, 262-272
canonical variation, 105
cardinality, 130
cargo, 250
Cartesian argument, 20
Cartesian product, 130, 152
Cartesian rationalism, 13
causation, 13
cause-effect, 155
CDF, 193
    probability, 193
    addition, 196
    arithmetic, 196
    bounds, 193
    division, 197
    interval, 189
    left, 195
    multiplication, 197
    right, 195
    subtraction, 197
central tendency, 116, 173
certain event, 129

Index

certainty equivalent, 112
chains, 199
Challenger shuttle, 103
Chamber, 251
chance node, 60
characteristic function, 128, 132
Christian, 12
Christianity, 16
chunking, 95
civil defense, 100
classical sets, 86
classification, 127
    failure, 168
    monotone measure, 164
closed-ended, 120
coarseness, 31-32, 34, 86, 90, 157
coefficient of variation, 177
cognition, 22, 89
cognitive knowledge, 21
cognitive science, 22
collapse, 227
collective scoring, 209
collectively exhaustive, 130
combination, 214
    opinion, 246
combinations of laws, 131
Commentator, 12
communication, 247
commutative laws, 131
comparability, 115, 118, 132
complement, 146, 154
    intensity factor, 146
    Sugeno class, 146
    Yager class, 146
laws, 131
complete failure, 150
complex systems, 55
complexity, 78, 91
    organized, 92
component, 58
component integration method, 58
composition, 154
Comte, 15-16
concept knowledge, 18
concrete, 190
confidence, 114
confidence bounds, 114
conflict, 31-33, 86
confusion, 31-33, 86, 213, 223
conjecture, 21, 23
conjunction, 199
connective, 144-145
conorms, 144
conscious ignorance, 28, 31-33
consciousness, 5

consensus, 109, 214-215, 239
consequence, 104, 256, 262-272
constraint, 141
construction, 69, 78, 81, 201
construction activity, 69
construction system, 81
construct-related evidence, 115
content, 262-272
content-related evidence, 115
content-to-structure value, 262-272
continuous set, 127
control, 106, 110, 155
conviction, 98
convolution, 196
cooperative action, 163
correlation, 116
cosmogony, 4
cosmology, 11
COV, 177
covariance, 177
cranes, 201
credibility, 77
crisis, 100
crisp, 68, 77
crisp partition, 157
crisp set, 127
criterion, 155
criterion-related evidence, 115
CSVR, 262-272
cumulative distribution function, see CDF
customer, 50

D

dam, 38, 43-44, 251
dam functions, 43
damage, 152, 227, 262-272
damage ratio, 262-272
damage spectrum, 152
damage-depth ratio, 262-272
Darwin, 16
data, 72, 107
data dependence, 107
data spread, 107
data system, 72
de Morgan's law, 131
decision, 48, 59
    arrow, 60
    chance node, 60
    consequences, 61
    deterministic node, 60
    method, 59
    node, 60
    outcomes, 61

probabilities, 61
symbols, 60
trees, 61
variables, 59
decision analysis, 59
decision maker, 97
deduction, 9
defective, 183
Defense Intelligence Agency (DIA), 109
definitions, 6
Delphi, 99, 113-114
Delphi questionnaire, 100
Dempster rule of combination, 167
Dempster's rule, 222
Dempster-Shafer monotone measure, 164
Department of Defense (DoD), 49, 104
dependencies, 196
dependent, 155
Descartes, 13-14
design, 47
design concepts, 47
design decision, 48
design details, 48
design selection, 47
detailed design, 48
development of issues, 243
Dewey, 16, 19
DIA, see Defense Intelligence Agency
dialectic, 6
dialectical laws, 16
dialectical materialism, 16
dialectical process, 15
dianoi, 21, 23
difference, 146
direct method, 113
discovery, 105
discrepancy measure, 212
discrete set, 127
discussion, 247
disjunction, 199
dispersion, 116, 176
dissonance, 213, 223
distribution selection, 219
distributive laws, 131
division, 140
documentation, 247
DoD, see Department of Defense
double-truth doctrine, 12
dualism, 13

E

Earth, 12
eikasia, 21, 23, 97

Einstein, 38
elevators, 250
elicitation, 99, 111, 234
 applications, 250
 calibration, 111
 indirect, 111
 issue, 111
 methods, 111
 opinions, 245
 question, 111
Empedocles, 6
empirical, 12
empirical control, 110
empirical sciences, 16
empiricism, 4-5, 13, 14-15, 17
empty set, 127
Emsworth, 251
Engels, 15-16
engineering systems, 49, 85
Enneads, 11
entropy, 211, 223
entropy-like, 211
environment, 39
Epictetus, 10
Epicurus, 10
episteme, 21, 23
epistemological, 7
epistemology, 3-5, 17
equating, 115, 118
equivalent, 112
error, 14-15, 94, 214, 215
essence, 13
estimate of probability, 172
ethics, 4, 11
Euclid, 6
Euclidean space, 129
evaluators, 235, 236
evidence, 20, 86
 monotone measure, 164
 probability, 170
evidence theory, 213
 basic assignment, 166
 belief, 165
 belief measure, 165
 null set, 165
 plausibility, 166
 plausibility measure, 166
 power set, 165-166
evolution theory, 5, 16
evolutionary epistemology, 18, 20
evolutionary infallible knowledge, 27
evolutionary knowledge, 98
evolutionary process, 17
excellent, 70
exemplification, 68

Index

existence, 13
expected utility, 112
experience, 70, 133, 153
experience models, 70
experimental naturalism, 17
experimentation, 17
expert, 27, 98, 126, 204, 234-236, 241
    training, 245
expert opinion, 98-99, 114, 234
    combination, 214
    consensus, 215
    error, 215
    generalized weighted average, 217
    groups, 215
    interval, 214
    maximum value, 217
    minimum value, 217
    norm probability, 216
    percentiles, 215
    scoring, 208
    uncertainty criteria, 218
    weighted, 215
    weighted arithmetic average, 216
    weighted geometric mean, 216
    weighted harmonic mean, 216
expert opinion aggregation, 231
expert opinion elicitation, 114
expert opinion elicitation process, 234-236, 238, 240
exponential, 220
extension, 13

F

failure, 227
    bridge, 168
    case, 168
    classification, 168, 227
    definition, 150, 222
    level, 151, 222
    rate, 173
fairness, 111
fallacy, 31-33, 91
familiarity knowledge, 18
fault tree, 109
Feather River, 262-272
filtering, 94
finite deterrence, 100
finite set, 127
first strike, 100
flood, 38, 56, 262-272
    extreme, 56
    maximum, 56
    PMF, 56
    probable maximum, 56
    runoff, 57
    standard project flood (SPF), 56
    control, 38
floodwater, 38, 262-272
focal element, 167
focus groups, 123
    Internet, 124
forecasting, 161
form, 7, 10
format, 121
frame, 262-272
Frechet, 199
frequency probability, 171
function, 76, 155
    alpha-cut, 155
    criterion, 155
    dependent, 155
    independent, 155
    predictor, 155
    relation, 155
    rough, 160
    triplet, 155
    variable, 155
functional requirements, 43, 46
funnel questions, 122
fuzzified evidence, 214
fuzziness, 214, 231
fuzzy, 68-69, 76, 85
    addition, 140
    alpha cut, 134
    arithmetic, 140
    boundaries, 135
    cause-effect, 155
    complement, 146
    connective, 145
    conorms, 145
    difference, 146
    division, 140
    function, 155
    intensity factor, 145-146
    intersection, 145
    measure, 162
    measure theory, 155
    monotone measure, 164
    multiplication, 140
    norms, 145
    operations, 131, 134, 143, 145
    power, 155
    power consumption, 155
    subset, 132
    subtraction, 140
    triangular conorms (t-conorms), 145
    triangular norms (t-norms), 145
    union, 144

Venn-Euler, 135
Yager class, 145-146
fuzzy arithmetic, 231
fuzzy control, 155
fuzzy entropy, 214
fuzzy event, 150
fuzzy failure, 150
fuzzy interval, 136
fuzzy logic, 155
fuzzy numbers, 136
    bounds, 136
    left, 136
    right lower, 136
    trapezoidal, 136
    triangular, 136
    upper, 136
fuzzy relation, 152
fuzzy set, 86, 127, 131
    core, 135
    nested, 135
    normal, 135
    quartile support, 135
    subnormal, 135
    support, 135
fuzzy rough, 157

G

Galileo, 12-13
general variable, 71
generalized means, 147
    aggregation, 147
    averaging, 147
    intensity factor, 147
generalized measure, 214
generalized weighted average, 217
generative, 75
generative behavior function, 76
generative system, 75, 81
generative uncertainty, 78
Gettier problem, 20, 22
GNP, 103
goal, 46
God, 6, 13
Gödel, 25, 31-33
good, 70
Gorgias, 6
gravitation, 12
Greek, 10
gross national product, 103
groups, 123, 214-215, 247
    focus, 123
    Internet, 124

H

happiness, 6
hard operations, 147
Hartley, 210
Hartley measure, 210, 230
Hartley-like, 210
heavens, 12
Hegel, 15-16
Heisenberg, 38
Hellenistic, 10
Heraclitus, 6
heuristics, 105
    anchoring, 106
    availability, 106
    Bayes, 107
    bias, 107
    control, 106
    overconfidence, 107
    representativeness, 106
high, 133
houses, 262-272
human factors, 87
human vision, 94
Hume, 14-15
Husserl, 16, 19
hypotheses, 21, 23
hypothetical situation, 105

I

Ibn Rushd, 5, 11-13
Ibn Sina, 5, 11-13
idea, 7, 10, 12
idealism, 5, 7, 14-15
idempotent laws, 131
identification of
    experts, 241
    issues, 243
    leader, 241
    peer reviewers, 241
identity laws, 131
ignorance, 6, 26-27, 83, 88
ignorance types, 86
ignoratio elenchi, 29
image, 8, 67
immune deficiency syndrome, 91
inaccuracy, 31-33, 86
incompleteness, 31-33
inconsistency, 31-33
independent, 155
indirect elicitation, 111
induction, 6
inductive inference, 12

Index

infallible evidence, 20
inference, 12
infinite set, 127
influence diagrams, 64
information, 18, 22-23, 66, 109, 204
    communication, 109
    intelligence, 109
    reliability, 109
    security, 204
    systems, 159
information overload, 94
inspection, 62
inspection strategies, 62
instrumentalism, 17
intelligence, 22, 109
intensity factor, 145-147
interaction, 247
Internet, 204
    focus groups, 124
    security, 204
interpretation, 68
intersection, 130, 145, 154
    connective, 146
    intensity factor, 146
    norms, 146
    triangular norms (t-norms), 146
    Yager class, 146
Interval, 136, 214
interval analysis, 86
interval approximation, 136
interval cumulative distribution function, 189
interval probabilities, 86
interval probability, 188
intrinsicism, 7
intruder, 204
invariance uncertainty criterion, 218
invisible, 8
irrationalism, 17
irrelevance, 31-33, 91
Islamic, 10, 12
Islamic Neoplatonism, 11
issue, 111, 236-237, 243

J

Jewish, 12
judge, 204
judgment, 98, 204
    probability, 171
justifiable true beliefs, 27
justification, 105
justified true belief, 20

K

Kant, 14-15
Kent chart, 109
know-how ignorance, 18, 29
knowledge, 3-4, 9, 18, 66
    evolutionary, 98
    speculative, 97
Kolmogorov-Smirnov, 191
kosmos aisthetos, 6
kosmos noetos, 6
KS method, 191

L

labor experience, 133
Lagrange multipliers, 220, 231
late-stage peer review, 236, 237
law, 204
leader of process, 235-236
left cumulative distribution function, 195
left bounds, 136
legal, 204
legal opinion, 204
level, 68
lifecycle, 49
likelihood, 31-32, 34, 78, 104
linguistics, 24
    probability, 173
living systems, 94
lock, 251
Locke, 13-14
logic, 9
logical atomism, 18
logical positivism, 17, 19
logical positivists, 17
long experience, 133
low, 133
lower approximation, 157

M

manufacturing, 183
mapping, 71, 163
Markov, 57
    transition, 57
Marx, 15-16
mask, 76
materialism, 5
materialistic monism, 5
materials, 201
mathematics, 21, 23
matrix, 153

matter, 4
max, 153
maximum uncertainty criterion, 218
maximum entropy, 220
maximum value, 217
mean, 173
    uncertainty, 191
measure, 209
    fuzzy, 162
    monotone, 162
    types, 209
measure theory, 155, 162
mechanics, 12
median, 173, 176
Medieval, 10
medium, 133
membership, 127, 152
membership function, 128, 132, 159, 214
memoryless, 76
meta-ignorance, 29
metaphysics, 3-4
metasystem, 82
methodological distinctions, 76
military, 123
min, 153
minimum uncertainty criterion, 218
minimum complexity, 78
minimum value, 217
mirrors, 8
mission, 48
miter gate, 256
mode, 13
model, 116
modeling, 85
moderate, 70
money, 111
monotone measure, 86, 162
    additive, 165
    belief, 164
    classification, 164
    Dempster-Shafer, 164
    evidence, 164
    fuzzy, 164
    necessity, 164
    nonadditive, 165
    plausibility, 164
    possibility, 164
    probability, 164
    rough, 165
monotonic behavior, 129
monotonousness, 129
Montaigne, 12-13
mooring, 256
motion, 12
multiplication, 130, 140

Muslim, 12
mutually exclusive, 130

N

National Aeronautics and Space Administration (NASA), 49, 51, 103
National Security Agency (NSA), 109
Naturalism, 17
nature, 14
naval, 250
navigation, 251
necessity, 77, 199
    monotone measure, 164
need identification, 50, 239
needs, 39
Neoplatonism, 9-10
nested sets, 135, 199
neuroscience, 24
neutrality, 111
Newton, 12-13, 76
Nietzsche, 15-16
nihilism, 17
nominal group technique, 113
nonadditive monotone measure, 163, 165
nonbelonging, 128, 132
nonconstraint, 141
nonfuzzy set, 127
noninteractive, 164
nonpropositional knowledge, 18
non-reflective ignorance, 29
nonrestricted, 140
nonspecificity, 31-33, 209
norm probability, 216
normal, 220
normal distribution, 114, 190
normal fuzzy set, 135
norming, 115, 118
norms, 145
null set evidence theory, 165
null set, 127
number approximation, 136
numbers, 136

O

object ignorance, 29
object knowledge, 18
object system, 67
objective, 46, 181
observation, 17, 68
observation channel, 71
observers, 235-236
Ohio River, 251

Index

omission, 94
ontological, 7
ontology, 4
open-ended, 120
operations, 130, 143, 157
   aggregation, 147
   averaging, 147
   generalized means, 147
   hard, 147
   intensity factor, 147
   soft, 147
opinion, 18, 22-23, 98, 126, 204
   aggregation, 246
   combination, 214, 246
   consensus, 215
   error, 215
   generalized weighted average, 217
   groups, 215
   interval, 214
   maximum value, 217
   minimum value, 217
   norm probability, 216
   percentiles, 215
   uncertainty criteria, 218
   weighted, 215
   weighted arithmetic average, 216
   weighted geometric mean, 216
   weighted harmonic mean, 216
opinion aggregation, 221
opinion elicitation, 99, 245
opinion scoring, 208
ordered, 67
ordered pairs, 152
ordering, 77
organized complexity, 92
outcomes, 236, 237

P

pairs, 152
parallelism, 13
parameters, 190
parametric estimation, 114
partial collapse, 227
partial ordering, 77
participatory peer review, 236-237
partition, 157
pattern recognition, 92
peer reviewers, 241, 235-236
   late-stage, 236-237
   participatory, 236-237
Peirce, 16, 19
people, 39
percentile, 177, 179, 214-215

performance, 149, 151
performance of mission, 48
performance requirements, 44
personal flotation device, 63
persuasion, 98
PFD, 63
phenomenalism, 14-15
phenomenology, 17, 19
philosophy, 3
physical randomness, 87
physical requirements, 44
picture theory of meaning, 17
pistis, 21, 23, 97
Plato, 7-10
Platonic, 9
Platonic love, 7
Platonism, 7, 10
plausibility, 77, 199, 213
   evidence theory, 166
   monotone measure, 164
   probability, 170
plausibility measure in evidence theory, 166
Plotinus, 9-10, 13
Poisson, 220
policy maker, 97
polyhedron, 129
poor, 70
population, 116, 178
positivism, 16, 119
possibility, 77, 86
   belief, 199
   chains, 199
   monotone measure, 164
   necessity, 199
   nested sets, 199
   plausibility, 199
   theory, 199
posterior, 182
posteriori, 4-5
posteriori principles, 13
power, 155
   forecasting, 161
power consumption, 155
power set in evidence theory, 165-166
power set, 130
pragmatism, 17, 19
precision, 94
predictor, 155
preSocratic, 6
prior, 182
priori, 4-5, 14
probability, 77, 86, 182
   additive, 170
   axiom, 172
   basic assignment, 171

belief, 170
bounds, 193
definition, 171
estimate, 172
evidence, 170
frequency, 171
interval, 188
judgment, 171
linguistic, 173
monotone measure, 164
plausibility, 170
set, 170
singleton, 170
subjective, 171
subset, 170
theory, 170
true, 172
probability bounds, 193
probability mass, 211
probability theory, 128, 212
process, 45
communication, 247
documentation, 247
process definition, 238
process modeling, 45
process outcomes, 236, 237
processing, 94
producer, 50
production, 183
Professional Standards for Test Use, 115
programming, 116
proof, 12
proponents, 235, 236
propositional ignorance, 29
propositional knowledge, 18
Protagoras, 7, 10
psychology, 22
publication, 115
Pyrrho, 10

Q

qualification, 116
quality, 153, 183
quartile, 136
question, 111, 120, 123
closed-ended, 120
format, 121
funnel, 122
open-ended, 120
scale, 121
units, 121
questionnaire, 100, 122
queuing, 94

R

rainfall, 262-272
RAND, 99, 114
randomness, 31-32, 34, 86
rate, 173
rational consensus, 109
rationalism, 4-5, 13-15
rationalist, 15
real system, 68
reality, 3, 116
reality theory, 5
reasoning, 21, 23
reflective ignorance, 29
relation, 152, 155
Cartesian product, 152
complement, 154
composition, 154
experience, 153
fuzzy, 152
intersection, 154
matrix, 153
max, 153
membership, 152
min, 153
ordered pairs, 152
pairs, 152
quality, 153
support, 153
union, 154
relevance, 94
reliability, 115, 149
reliability rating, 109
reliability theory of knowledge, 21
reliable knowledge, 27
religious body, 12
Renaissance, 12
representativeness, 106
reproducibility, 107, 109
requirement analysis, 43
requirements, 43
resource experts, 235-236
response, 150
restricted, 140
retina, 93
revelation, 12
reverse probability, 182
right cumulative distribution function, 195
right bounds, 136
risk analysis, 234
risk assessment, 104
risk index, 104
river, 57
rough, 85
rough monotone measure, 165

Index

rough function, 160
rough sets, 86, 90
    coarseness, 157
    crisp partition, 157
    fuzzy rough, 157
    information, 159
    lower approximation, 157
    membership function, 159
    operations, 157
    partition, 157
    rough fuzzy, 157
    universal space, 157
    upper approximation, 157
rough fuzzy, 157
Royce, 16, 19
rule of combination, 167, 222
Russell, 16, 19

S

safety, 38
sample, 178
sample size, 116
sample space, 128
sampling, 31-32, 34, 86
Santa Fe Institute, 91
scale, 121
scaling, 115, 118
scenario, 105
    analysis, 105
    surprise-free, 105
science, 105
scientific body, 12
scientific heuristics, 105
scientific verification, 17
scoring, 208
    collective, 209
    self, 208
security, 204
selection of experts, 241
selection of peer reviewers, 241
self disclosure, 123
self scoring, 208
semicontinuous, 163
sentiment, 98
SEPT, 114
    primary standards, 115
    Professional Standards for Test Use, 115
    secondary standards, 115
    Standards for Administrative Procedures, 115
    Standards for Particular Applications, 115
    technical standards, 115
    Technical Standards for Test Construction and Evaluation, 115
serviceability, 227
serviceability failure, 150
set, 127
    cardinality, 130
    classification, 127
    continuous, 127
    convex, 128
    core, 135
    crisp, 127
    discrete, 127
    empty, 127
    family, 163
    finite, 127
    fuzzy, 127, 131
    infinite, 127
    member, 128
    membership, 127
    nested, 135
    nonconvex, 128
    nonfuzzy, 127
    null, 127
    power, 130
    probability, 170
    quartile support, 135
    rough, 156
    subset, 128
    support, 135
set operations, 130-131
set theory, 127-128
shadows, 8
Shannon, 211
Shannon entropy, 212, 223, 230
sharp, 136
ships, 250
short experience, 133
siding, 262-272
simplification, 31-32, 34, 86
simulation, 116
singleton probability, 170
skepticism, 10, 12
snow, 262-272
social Darwinism, 16
social phenomena, 119
social research methods, 119
social research stages, 119
social research, 118
sociology of ignorance, 26
sociology of knowledge, 26
sociology of nonknowledge, 26
Socrates, 6
soft operations, 147
soul, 11
source system, 67, 69

specific image system, 67
specific variable, 71
speculative knowledge, 97
speech-act theory, 18
sphere, 129
Spinoza, 13-14
sponsor of process, 235, 236
spread, 116
standard deviation, 177
    uncertainty, 191
standard error, 116
standard normal, 114
Standards for Administrative Procedures, 115
Standards for Educational and Psychological
    Testing, see SEPT
Standards for Particular Applications, 115
state, 57
state table, 80
state-based method, 57
state-transition function, 76, 80
statistical uncertainty, 87, 178
statistics, 86
storm, 38, 262-272
strength, 190
strong alpha cut, 134
strongest, 167
structural reliability, 149
structural response, 150
structural system, 81
structure, 58, 150, 227, 262-272
study complexity, 236, 237
study leader, 240
study level, 236-237, 240
subject, 107, 122, 235-236
    calibration, 107
subjective, 182
subjective probability, 171
subnormal fuzzy set, 135
subset, 132
subset for probability, 170
substance, 13-14
subtraction, 130, 140
Sugeno class, 146
support, 153
support set, 70
surprise-free scenario, 105
survival, 150
synergy, 163
synthesis, 15
synthetic posteriori, 15
synthetic priori, 15
system, 38, 57
    abstraction, 68
    advanced studies, 51
    attributes, 68

backdrop, 67
certification, 53
conceptual definition, 52
conceptual design, 52
construction, 50
control, 66
data, 72
definition, 66
design, 50, 53
development, 53
disposal, 54
evaluation, 50
exemplification, 68
fabrication, 53
general image, 67
generative, 75
hierarchy, 66
image, 67
information, 66
integration, 53
interpretation, 68
knowledge, 66
level, 68
living, 94
logistic support, 50
object, 67
observation, 68
operation, 53, 54
planning, 50
production, 50
real, 68
repairable, 57
research, 50
source, 67
specific image, 67
structure, 58
test, 53
truss, 58
use, 50
system analysis, 42
system complexity, 91
system definition, 42
system engineer, 38
system engineering, 46
system engineering process, 46
system goal, 46
system objectives, 46
system science, 42
system types, 42

T

taboo, 31-33
t-conorms, see triangular conorms

Index

technical facilitator, 235-236, 240
technical integrator, 235-236, 240
technical integrator and facilitator, 235-236, 240
technical issues, 243
technical maturity model, 54
Technical Standards for Test Construction and Evaluation, 115
test development, 115
testing, 47
Thales, 3
theory of being, 7
theory of evolution, 5, 16
theory of forms, 7
theory of ignorance, 26
theory of knowledge, 7
theory of reality, 5, 7, 10
theory of relativity, 25
thermonuclear war, 102
thesis, 15
thought, 13
ticket, 112
time, 25
time asymmetry, 25
time to failure, 176
t-norms, see triangular norms
total expected cost, 62
total probability theorem, 182
tow, 256
tower cranes, 201
tradeoff, 47, 94
traffic, 202
training of experts, 245
transcomputational problems, 92
transition, 57
trapezoidal fuzzy number, 136
triangular conorms, 144
triangular fuzzy number, 136
triangular norms, 145
Trinity, 12
triplet, 155
true probability, 172
truss, 58
truth, 11-12, 15
Tuhafut al-Falasefah, 12
Tuhafut al-Tuhafut, 12

U

uncertainty, 31-33, 78, 83
   mean, 191
   measure, 209
   parameter, 190
   population, 178
   sample, 178
   standard deviation, 191
   statistical, 178
   types of measure, 209
uncertainty criteria for expert opinion, 218
   invariance, 218
   maximum, 218
   minimum, 218
uncertainty invariance, 221
uncertainty measures, 209
   assignment, 211, 213
   belief, 213
   binary, 210
   bits, 210
   byte, 210
   confusion, 213
   discrepancy measure, 212
   dissonance, 213
   entropy, 211
   entropy-like, 211
   estimate, 212
   evidence theory, 213
   fuzzified evidence, 214
   fuzziness, 214
   fuzzy entropy, 214
   fuzzy set, 213
   generalized measure, 214
   Hartley, 210
   Hartley measure, 210
   Hartley-like, 210
   membership function, 214
   nonspecificity, 209
   plausibility, 213
   probability mass, 211
   probability theory, 212
   Shannon, 211
   Shannon entropy, 211
   true value, 212
   U-uncertainty, 211
undecidability, 31-33
uniform, 220
union, 130, 144, 154
   connective, 144
   conorms, 144
   intensity factor, 144
   operations, 144
   triangular conorms, 144
   Yager class, 144
universal space, 157
universe, 128, 132
unknowable, 31-33
unknown, 90, 31-33
unspecificity, 31-33
untopicality, 31-33
upper approximation, 157
USACE, see Army Corps of Engineers

USSR, 109
utility, 112
utility value, 111
U-uncertainty, 211

V

vagueness, 31-32, 34, 86-87
validation, 47, 116
validity, 115
validity generalization, 116
variability, 176
variable, 67, 155
variance, 116, 177
vehicle, 262-272
vehicle damage, 262-272
Venn, 129
Venn-Euler diagram, 129
    fuzzy, 135
verifiability, 18
verification, 116
vertex method, 140
Vienna Circle, 18
view, 98

visible, 8

W

war, 102
War Department, 123
weakest, 167
weighing, 239
weighted, 214-215
weighted arithmetic average, 216
weighted geometric mean, 216
weighted harmonic mean, 216
wildlife, 204
Wittgenstein, 16, 19
wood, 262-272
work breakdown, 43, 45
worker experience, 133

Y

Yager class, 144-145
Yuba County, 262-272

T 56 .A98 2001
Ayyub, Bilal M.
Elicitation of expert
opinions for uncertainty